DICTIONARY OF
APPLIED MATH FOR ENGINEERS AND SCIENTISTS

COMPREHENSIVE DICTIONARY
OF MATHEMATICS

Douglas N. Clark
Editor-in-Chief

Stan Gibilisco
Editorial Advisor

PUBLISHED VOLUMES

Analysis, Calculus, and Differential Equations
Douglas N. Clark

Algebra, Arithmetic, and Trigonometry
Steven G. Krantz

Classical and Theoretical Mathematics
Catherine Cavagnaro and William T. Haight, II

Applied Mathematics for Engineers and Scientists
Emma Previato

FORTHCOMING VOLUMES

The Comprehensive Dictionary of Mathematics
Douglas N. Clark

A VOLUME IN THE
COMPREHENSIVE DICTIONARY
OF MATHEMATICS

DICTIONARY OF

APPLIED MATH FOR ENGINEERS AND SCIENTISTS

Edited by

Emma Previato

CRC PRESS

Boca Raton   London   New York   Washington, D.C.

Library of Congress Cataloging-in-Publication Data

Dictionary of applied math for engineers and scientists/ edited by Emma Previato.
   p. cm.
  ISBN 1-58488-053-8
   1. Mathematics—Dictionaries. I. Previato, Emma.

QA5 .D49835 2002
510′.3—dc21   2002074025

Visit the CRC Press Web site at www.crcpress.com

PREFACE

To describe the scope of this work, I must go back to when Stan Gibilisco, editorial advisor of the dictionary series, asked me to be in charge of this volume. I appreciated the idea of a compendium of mathematical terms used in the sciences and engineering for two reasons. Firstly, mathematical definitions are not easily located; when I need insight on a technical term, I turn to the analytic index of a monograph that seems related; recently I was at a loss when trying to find "Viète's formulas *," a term used by an Eastern-European student in his homework. I finally located it in the *Encyclopaedic Dictionary of Mathematics*, and that brought home the value of a collection of esoteric terms, put together by many people acquainted with different sectors of the literature. Secondly, at this time we do not yet have a tradition of cross-disciplinary terms; in fact, much interaction between mathematics and other scientific areas is in the making, and times (and timing) could not be more exciting. The EPSRC** newsletter Newsline (available on the web at www.epsrc.ac.uk), devoted to mathematics, in July 2001 rightly states "Even amongst fellow scientists, mathematicians are often viewed with suspicion as being interested in problems far removed from the real world. But . . . things are changing."

Rapidly, though, my enthusiasm turned to dismay upon realizing that any strategy I could devise was doomed to fail the test of "completeness." What is a dictionary? At best, a rapidly superseded record of word/symbol usage by some groups of people; the only really complete achievement in that respect is, in my view, the OED. Not only was such an undertaking beyond me, the very attempt at bridging disciplines and importing words from one to another is still an ill-defined endeavor — scientists themselves are unsure how to translate a term into other disciplines.

As a consequence what service I can hope this book to provide, at best, is that of a pocket manual with which a voyager can at least get by in a basic fashion in a foreign-speaking country. I also hope that it will have the small virtue to be a first of its kind, a path-breaker that will prompt others to follow. Not being an applied mathematician myself, I relied on the generosity of the following team of authors: Lorenzo Fatibene, Mauro Francaviglia, and Rudolf Schmid, experts of mathematical physics; Toni Kazic, a biologist with broad and daring interdisciplinary experience; Hong Qian, a mathematical biologist; and Ralf Hiptmair, who works on numerical solution of differential equations. For operations research, Giovanni Andreatta (University of Padua, Italy), directed me to H.J. Greenberg's web glossary, and Toni Kazic referred me to the most extensive web glossary in chemistry, authored by A.D. McNaught and A. Wilkinson. To all these people I owe much more than thanks for their work. I know the reward that would most please them is for this book to have served its readers well: please write me any comments or suggestions, and I will gratefully try to put them to future use.

Emma Previato, Department of Mathematics and Statistics
Boston University, Boston, MA 02215-2411 – USA
e-mail: ep@bu.edu

---

*They are just the elementary symmetric polynomials, in case anyone beside me didn't know
**Engineering and Physical Sciences Research Council, UK.

CONTRIBUTORS

Lorenzo Fatibene
Istituto di Fisica Matematica
Università di Torino
Torino, Italy

Mauro Francaviglia
Istituto di Fisica Matematica
Università di Torino
Torino, Italy

Ralf Hiptmair
Mathematisches Institut
Universität Tübingen
Tübingen, Germany

Toni Kazic
Department of Computer Engineering and
    Computer Science
University of Missouri — Columbia
Columbia, Missouri, U.S.

Hong Qian
Department of Applied Mathematics
University of Washington
Seattle, Washington, U.S.

Rudolf Schmid
Department of Mathematics and
    Computer Science
Emory University
Atlanta, Georgia, U.S.

To Professor Greenberg and Dr. McNaught, a great many thanks are due for a most courteous, prompt and generous permission to use of their glossaries.

Harvey J. Greenberg
Mathematics Department
University of Colorado at Denver
Denver, Colorado, U.S.

A.D. McNaught
General Manager, Production Division
RSC Publishing, Royal Society of Chemistry
Cambridge, U.K.

In addition, the two following databases have been used with permission:
IUPAC Compendium of Chemical Terminology, 2nd ed. (1997), compiled by Alan D. McNaught and Andrew Wilkinson, Royal Society of Chemistry, Cambridge, U.K. http://www.iupac.org/publications/compendium/index.html
H. J. Greenberg. Mathematical Programming Glossary http://carbon.cudenver.edu/˜hgreenbe/glossary/ glossary.html ,1996-2000.

A

a posteriori error estimator   An algorithm for obtaining information about a discretization error for a concrete discrete approximation u_h of the continuous solution u. Two principal features are expected from such device:

(i.) It should be *reliable*: the estimated error (norm) must be proportional to an upper bound for the true error (norm). Thus, discrete solutions that do not meet a prescribed accuracy can be detected.

(ii.) It should be *efficient*: the error estimator should provide some lower bound for the true error (norm). This helps avoid rejecting a discrete solution needlessly.

In the case of a finite element discretization an additional requirement is the *locality* of the a posteriori error estimator. It must be possible to extract information about the contributions from individual cells of the mesh to the total error. This is essential for the use of an a posteriori error estimator in the framework of *adaptive refinement*.

abacus   Oldest known "computer" circa 1100 BC from China, a frame with sliding beads for doing arithmetic.

Abbe's sine condition   (Ernst Abbe 1840–1905) $n'l' \sin \beta' = nl \sin \beta$ where n, n', β, β' are the refraction indices and refraction angles, respectively.

Abelian group   (Niels Henrik Abel 1802–1829) A *group* $(G, \cdot)$ is called Abelian or commutative if $a \cdot b = b \cdot a$ for all $a, b \in G$.

Abelian theorems   (1) Suppose $\sum_{n=0}^{\infty} a_n x^n$ converges for $|x| < R$ and for $x = R$. Then the series converges uniformly on $0 \leq x \leq R$.
(2) For $n \geq 5$ the general equation of nth order cannot be solved by radicals.

Abel's integral equation $f(x) = \int_0^x \frac{\phi(\xi)}{\sqrt{x-\xi}} d\xi$, where $f(x)$ is C^1 with $f(0) = 0$, is called *Abel's integral equation*.

aberration   The deviation of a spherical mirror from perfect focusing.

abscissa   In a rectangular coordinate system (Cartesian coordinates) (x, y) of the plane $\mathbf{R}^2$, x is called the *abscissa*, y the ordinate.

absolute convergence   A series $\sum x_n$ is said to be *absolute convergent* if the series of *absolute values* $\sum |x_n|$ converges.

absolute convergence test   If $\sum |x_n|$ converges, then $\sum x_n$ converges.

absolute error   The difference between the exact value of a number x and an approximate value a is called the *absolute error* Δ_a of the approximate value, i.e., $\Delta_a = |x - a|$. The quotient $\delta_a = \frac{\Delta_a}{a}$ is called the relative error.

absolute ratio test   Let $\sum x_n$ be a series of nonnegative terms and suppose $\lim_{n \to \infty} \frac{|x_{n+1}|}{|x_n|} = \rho$.

(i.) If $\rho < 1$, the series converges absolutely (hence converges);

(ii.) If $\rho > 1$, the series diverges;

(iii.) If $\rho = 1$, the test is inconclusive.

absolute temperature $-273.15°C$.

absolute value   The *absolute value* of a real number x, denoted by $|x|$, is defined by $|x| = x$ if $x \geq 0$ and $|x| = -x$ if $x < 0$.

absolute value of an operator   Let A be a *bounded linear operator* on a Hilbert space, $\mathbf{H}$. Then the *absolute value* of A is given by $|A| = \sqrt{A^* A}$, where A^* is the adjoint of A.

absolutely continuous   A function $x(t)$ defined on $[a, b]$ is called *absolutely continuous* on $[a, b]$ if there exists a function $y \in L^1[a, b]$ such that $x(t) = \int_a^t y(s)ds + C$, where C is a constant.

absorbance   A logarithm of the ratio of incident to transmitted *radiant power* through a sample (excluding the effects on cell walls). Depending on the base of the logarithm, decadic or Napierian absorbance are used. Symbols: A, A_{10}, A_e. This quantity is sometimes called extinction, although the term extinction, better called *attenuance*, is reserved for the quantity which takes into account the effects of luminescence and scattering as well.

absorbing set   A *convex set* $A \subset X$ in a *vector space* X is called *absorbing* if every $x \in X$ lies in tA for some $t = t(x) > 0$.

acceleration   The *rate of change* of *velocity* with time.

acceleration vector   If $\vec{v}$ is the *velocity* vector, then the *acceleration vector* is $\vec{a} = \frac{d\vec{v}}{dt}$; or if $\vec{s}$ is the *vector* specifying position relative to an origin, we have $\vec{v} = \frac{d\vec{s}}{dt}$ and hence $\vec{a} = \frac{d^2\vec{s}}{dt^2}$.

acceptor   A compound which forms a chemical bond with a substituent group in a bimolecular chemical or biochemical reaction.
     Comment: The donor-acceptor formalism is necessarily binary, but reflects the reality that few if any truly thermolecular reactions exist. The bonds are not limited to covalent. See also *donor*.

accumulation point   Let $\{z_n\}$ be a *sequence* of *complex numbers*. An *accumulation point* of $\{z_n\}$ is a complex number a such that, given any $\epsilon > 0$, there exist infinitely many integers n such that $|z_n - a| < \epsilon$.

accumulator   In a computing machine, an adder or counter that augments its stored number by each successive number it receives.

accuracy   Correctness, usually referring to numerical computations.

acidity function   Any function that measures the thermodynamic *hydron*-donating or -accepting ability of a solvent system, or a closely related thermodynamic property, such as the tendency of the *lyate ion* of the solvent system to form *Lewis adducts*. (The term "basicity function" is not in common use in connection with basic solutions.) Acidity functions are not unique properties of the solvent system alone, but depend on the solute (or family of closely related solutes) with respect to which the thermodynamic tendency is measured.
     Commonly used acidity functions refer to concentrated acidic or basic solutions. Acidity functions are usually established over a range of composition of such a system by UV/VIS spectrophotometric or NMR measurements of the degree of hydronation (protonation or *Lewis*

adduct formation) for the members of a series of structurally similar indicator bases (or acids) of different strengths. The best known of these functions is the Hammett acidity function H_0 (for uncharged indicator bases that are primary aromatic amines).

action   (1) The *action* of a conservative *dynamical system* is the space integral of the total *momentum* of the system, i.e.,

$$\int_{P_1}^{P_2} \sum_i m_i \frac{d\vec{r}_i}{dt} \cdot d\vec{r}_i$$

where m_i is the mass and $\vec{r}_i$ the position of the ith particle, t is time, and the system is assumed to pass from configuration P_1 to P_2.
     (2) *Action of a group:* A (left) *action* of a *group* G on a set M is a map $\Phi : G \times M \longrightarrow M$ such that:
     (i.) $\Phi(e, x) = x$, for all $x \in M$, e is the *identity* of G;
     (ii.) $\Phi(g, \Phi(h, x)) = \Phi(g \cdot h, x)$, for all $x \in M$ and $g, h \in G$. ($g \cdot h$ denotes the group operation (multiplication) in G.

If G is a *Lie group* and M is a smooth *manifold*, the action is called *smooth* if the map Φ is smooth.
     An action is said to be:
     (i.) *free* (without *fixed points*) if $\Phi(g, x) = x$, for some $x \in M$ implies $g = e$;
     (ii.) *effective* (faithful) if $\Phi(g, x) = x$ for all $x \in M$ implies $g = e$;
     (iii.) *transitive* if for every $x, y \in M$ there exists a $g \in G$ such that $\Phi(g, x) = y$.

See also *left action, right action*.

action angle coordinates   A system of generalized coordinates (Q_i, P_i) is called *action angle coordinates* for a *Hamiltonian system* defined by a Hamiltonian function H if H depends only on the generalized momenta P_i but not on the generalized positions Q_i. In these coordinates *Hamilton's equations* take the form

$$\frac{\partial P_i}{\partial t} = 0 \,, \quad \frac{\partial Q_i}{\partial t} = \frac{\partial H}{\partial P_i}$$

action, law of action and reaction (Newton's third law) The basic law of mechanics asserting that two particles interact so that the forces exerted by one upon another are equal in magnitude, act along the straight line joining the particles, and are opposite in direction.

action functional In variational calculus (and, in particular, in mechanics and in field theory) is a functional defined on some suitable space $\mathcal{F}$ of functions from a space of independent variables X to some target space Y; for any *regular domain* D and any configuration ψ of the system it associates a (real) number $A_D[\psi]$. A *regular domain* D is a subset of the space X (the time $t \in \mathbb{R}$ in mechanics and the space-time point $x \in M$ in field theory) such that the *action functional* is well-defined and finite; e.g., if X is a manifold, D can be any compact submanifold of X with a boundary ∂D which is also a compact submanifold.

By the *Hamilton principle*, the configurations ψ which are critical points of the action functional are called *critical configurations* (*motion curves* in mechanics and *field solutions* in field theory).

In mechanics one has $X = \mathbb{R}$ and the relevant space is the tangent bundle TQ to the *configuration manifold* Q of the system. Let $\hat{\gamma} = (\gamma, \dot{\gamma})$ be a *holonomic curve* in TQ which projects onto the curve γ in Q and $L : TQ \rightarrow \mathbb{R}$ be the *Lagrangian* of the system, i.e., a (real) function on the space TQ. The action is given by $A_D[\gamma] = \int_D L(\gamma(t), \dot{\gamma}(t)) \, dt$. D can be any closed interval. If suitable boundary conditions are required on γ one can allow also infinite intervals in the parameter space $\mathbb{R}$.

In field theory X is usually a *space-time* manifold M and the relevant space is the k-order jet extension $J^k B$ of the *configuration bundle* (B, M, π, F) of the system. Let $\hat{\sigma}$ be a *holonomic section* in $J^k B$ which projects onto the section σ in B and $L : J^k B \rightarrow \mathbb{R}$ be the *Lagrangian* of the system, i.e., a (real) function on the space $J^k B$. The action is given by $A_D[\sigma] = \int_D L(\hat{\sigma}(x)) \, ds$, where $L(\hat{\sigma}(x))$ denotes the value which the Lagrangian takes over the section; $D \subset M$ can be any regular domain and ds is a volume element. If suitable boundary conditions are required on the sections σ one can allow also infinite regions up to the whole parameter space M.

action principle (Newton's second law) Any force $\vec{F}$ acting on a body of mass m induces an acceleration $\vec{a}$ of that body, which is proportional to the force and in the same direction $\vec{F} = m\vec{a}$.

action, principle of least The principle (Maupertius 1698–1759) which states that the actual motion of a conservative dynamical system from P_1 to P_2 takes place in such a way that the *action* has a *stationary* value with respect to all possible paths between P_1 and P_2 corresponding to the same *energy* (*Hamilton principle*).

activation energy (Arrhenius activation energy) An empirical parameter characterizing the exponential temperature dependence of the *rate coefficient* k, $E_a = RT^2(d \ln k/dT)$, where R is the gas constant and T the thermodynamic temperature. The term is also used for *threshold energies* in electronic potential surfaces, in which case the term requires careful definition.

activity In biochemistry, the catalytic power of an enzyme. Usually this is the number of substrate turnovers per unit time.

adaptive refinement A strategy that aims to reduce some discretization error of a finite element scheme by repeated local refinement of the underlaying *mesh*. The goal is to achieve an equidistribution of the contribution of individual cells to the total error. To that end one relies on a local *a posteriori error estimator* that, for each cell K of the current mesh Ω_h, provides an estimate η_K of how much of the total error is due to K.

Starting with an initial mesh Ω_h, the refinement loop comprises the following stages:

(i.) Solve the problem discretized by means of a *finite element space* built on Ω_h;

(ii.) Determine guesses for the total error of the discrete solution and for the local error contributions η_h. If the total error is below a prescribed threshold, then terminate the loop;

(iii.) Mark those cells of Ω_h for refinement whose local error contributions are above the average error contribution;

(iv.) Create a new mesh by refining marked cells of Ω_h and go to (i.).

Algorithms for the local refinement of simplicial and hexaedral meshes are available.

addition reaction   A *chemical reaction* of two or more reacting *molecular entities*, resulting in a single reaction product containing all atoms of all components, with formation of two chemical bonds and a net reduction in bond multiplicity in at least one of the reactants. The reverse process is called an *elimination* reaction. If the reagent or the source of the addends of an addition are not specified, then it is called an addition transformation.

See also [*addition, α-addition, cheletropic reaction, cycloaddition.*]

adduct   A new *chemical species* AB, each *molecular entity* of which is formed by direct combination of two separate molecular entities A and B in such a way that there is change in *connectivity*, but no loss, of atoms within the moieties A and B. Stoichiometries other than 1:1 are also possible, e.g., a bis-adduct (2:1). An *intramolecular adduct* can be formed when A and B are groups contained within the same molecular entity.

This is a general term which, whenever appropriate, should be used in preference to the less explicit term *complex*. It is also used specifically for products of an *addition reaction*.

adiabatic lapse rate (in atmospheric chemistry) The rate of decrease in temperature with increase in altitude of an air parcel which is expanding slowly to a lower atmospheric pressure without exchange of heat; for a descending parcel it is the rate of increase in temperature with decrease in altitude. Theory predicts that for dry air it is equal to the acceleration of gravity divided by the specific heat of dry air at constant pressure (approximately $9.8°Ckm^{-1}$). The moist adiabatic lapse rate is less than the dry adiabatic lapse rate and depends on the moisture content of the air mass.

adjacency list   A list of edges of a graph G of the form

$$[v_i - [v_j, v_k, \ldots, v_n], v_j - [v_i, v_l, \ldots, v_m]],$$

$$\ldots, v_n - [v_i, v_p, \ldots v_q],$$

where

$$\mathcal{E} = \{(v_i, v_j), (v_i, v_k), \ldots, (v_i, v_n), (v_j, v_l),$$

$$\ldots, (v_j, v_m), \ldots, (v_n, v_p), \ldots, (v_n, v_q)\},$$

and $i, j, k, l, m, n, p,$ and q are indices.

Comment: Note that in this version any node is present at least twice: as the key to each sublist $(X - [\ldots]$ and as a member of some other sublist $(-[X])$. This representation is a more compact version of the connection tables often used to represent compound structures.

adjacent   For any graph $G(\mathcal{V}, \mathcal{E})$, two nodes v_i, $v_{i'}$ are *adjacent* if they are both incident to the same edge (share an edge); that is, if the edge $(v_i, v_{i'}) \in \mathcal{E}$. Similarly, two edges $(v_i, v_{i'})$, $(v_{i'}, v_{i''})$ are adjacent if they are both incident to the same vertex; that is if $\{v_i, v_{i'}\} \cap \{v_{i'}, v_{i''}\} \neq \emptyset$.

Comment: Two atoms are said to be *adjacent* if they share a bond; two *reactions (compounds)* are said to be *adjacent* if they share a *compound (reaction)*.

adjoint representations (on a [Lie] group G) (1) The *action* of any group G onto itself defined by ad : $G \rightarrow \text{Hom}(G) : g \mapsto \text{ad}_g$. The group automorphism $\text{ad}_g : G \rightarrow G$ is defined by $\text{ad}_g(h) = g \cdot h \cdot g^{-1}$.

(2) On a Lie algebra. If G is a *Lie group* the adjoint representation above induces by derivation the *adjoint representation* of G on its *Lie algebra* $\mathfrak{g}$. It is defined by $T_e \text{ad}_g : \mathfrak{g} \rightarrow \mathfrak{g}$ where T_e denotes the *tangent map* (see *tangent lift*). If G is a matrix group, then the adjoint representation is given by $T_e \text{ad}_g(\xi) = g \cdot \xi \cdot g^{-1}$.

(3) Also defined is the *adjoint representation* $\text{Ad} : \mathfrak{g} \rightarrow \text{Hom}(\mathfrak{g})$ of the Lie algebra $\mathfrak{g}$ onto itself. For $\xi, \zeta \in \mathfrak{g}$, the Lie algebra homomorphism $\text{Ad}_\xi : \mathfrak{g} \rightarrow \mathfrak{g}$ is defined by *commutators* $\text{Ad}_\xi(\zeta) = [\xi, \zeta]$.

adsorbent   A condensed phase at the surface of which *adsorption* may occur.

adsorption   An increase in the concentration of a dissolved substance at the interface of a condensed and a liquid phase due to the operation of surface forces. *Adsorption* can also occur at the interface of a condensed and a gaseous phase.

adsorptive   The material that is present in one or other (or both) of the bulk phases and capable of being adsorbed.

affine bundle   A bundle $(A, M, \pi; \mathbb{A})$ which has an affine space $\mathbb{A}$ as a standard fiber and *transition functions* acting on $\mathbb{A}$ by means of affine transformations.

If the base *manifold* is paracompact then any *affine bundle* allows global sections. Examples of *affine bundles* are the bundles of *connections* (transformation laws of connections are affine) and the *jet bundles* $\pi_k^{k+1} : J^{k+1}C \to J^kC$.

affine connection   A *connection* on the *frame* bundle $F(M)$ of a *manifold* M.

affine coordinates   An *affine coordinate* system $(0; x_1, x_2, ..., x_n)$ in an *affine space* $\mathbf{A}$ consists of a fixed pont $0 \in \mathbf{A}$, and a *basis* $\{x_i\}$ $(i = 1, ..., n)$ of the difference space E. Then every pont $P \in \mathbf{A}$ determines a system of n numbers $\{\xi^i\}$ $(i = 1, ..., n)$ by $\vec{0P} = \sum_{i=1}^n \xi^i x_i$. The numbers $\{\xi^i\}$ $(i = 1, ..., n)$ are called the *affine coordinates* of P relative to the given system. The origin 0 has coordinates $\xi_i = 0$.

affine equivalence   A special case of *parametric equivalence*, where the mapping Φ is an affine linear mapping, that is, $\Phi(\mathbf{x}) := A\mathbf{x} + \mathbf{t}$ with a regular matrix $A \in \mathbb{R}^{n,n}$ and $\mathbf{t} \in \mathbb{R}^n$ (n the dimension of the ambient space).

For affine equivalent families of *finite element spaces* on simplicial *meshes* in dimension n the usual reference element is the unit simplex spanned by the canonical basis vectors of $\mathbb{R}^n$.

In the case of a *shape regular* family $\{\Omega_h\}_{h \in \mathbb{H}}$ of meshes and *affine equivalence*, there exist constants $C_i > 0, i = 1, ..., 4$, such that

$$C_1 \text{diam}(K)^n \leq |\det A_K| \leq C_2 \text{diam}(K)^n,$$

$$\|A_K\| \leq C_3 \text{diam}(K),$$

$$\|A_K^{-1}\| \leq C_4 \text{diam}(K)^{-1} \; \forall K \in \Omega_h, h \in \mathbb{H}.$$

Here $\|.\|$ denotes the Euclidean matrix norm, and the matrix A belongs to that unique *affine mapping* taking a suitable reference element on K. These relationships pave the way for assessing the behavior of norms under pullback.

affine frame   An *affine frame* on a *manifold* M at $x \in M$ consists of a point $p \in A_x(M)$ (where $A_x(M)$ is the *affine space* with difference space $E = T_xM$) and an ordered basis $(X_1, ..., X_n)$ of T_xM (called a *linear frame at* x). It is denoted by $(p; X_1, ..., X_n)$.

affine geometry   The geometry of *affine spaces*.

affine map   Let X and Y be *vector spaces* and C a *convex* subset of X. A map $T : C \to Y$ is called an *affine map* if $T((1 - t)x + ty) = (1 - t)Tx + tTy$ for all $x, y \in C$, and all $0 \leq t \leq 1$.

affine mapping   (1) Let $\mathbf{A}$ be an *affine space* with difference space E. Let $P \mapsto P'$ be a mapping from $\mathbf{A}$ into itself subject to the following conditions:

(i.) $\vec{P_1Q_1} = \vec{P_2Q_2}$ implies $\vec{P_1'Q_1'} = \vec{P_2'Q_2'}$;

(ii.) The mapping $\phi : E \to E$ defined by $\phi(\vec{PQ}) = \vec{P'Q'}$ is linear.

Then $P \mapsto P'$ is called an *affine mapping*. If a fixed origin 0 is used in $\mathbf{A}$, every *affine mapping* $x \mapsto x'$ can be written in the form $x' = \phi x + b$, where ϕ is the induced linear mapping and $b = \vec{00'}$.

(2) Let M, M' be *Riemannian manifolds*. A map $f : M \to M'$ is called an *affine map* if the tangent map $Tf : TM \to TM'$ maps every horizontal curve into a horizontal curve. An *affine map* f maps every *geodesic* of M into a geodesic of M'.

affine representation   A representation of a *Lie group* G which operates on a *vector space* V such that all $\phi_g : V \to V$ are *affine maps*.

affine space   Let E be a real n-dimensional *vector space* and $\mathbf{A}$ a set of elements $P, Q, ...$ which will be called points. Assume that a relation between points and vectors is defined in the following way:

(i.) To every ordered pair (P,Q) of $\mathbf{A}$ there is an assigned vector of E, called the difference vector and denoted by $\vec{PQ}$;

(ii.) To every point $P \in \mathbf{A}$ and every vector $x \in E$ there exists exactly one point $Q \in \mathbf{A}$ such that $\vec{PQ} = x$;

(iii.) If P, Q, R are arbitrary points in $\mathbf{A}$, then $\vec{PQ} + \vec{QR} = \vec{PR}$.

Then $\mathbf{A}$ is called an *n-dimensional affine space* with *difference space* E.

The *affine n-dimensional space* A^n is distinguished from $\mathbb{R}^n$ in that there is no fixed origin; thus the sum of two points of A^n is not defined, but their difference is defined and is a vector in $\mathbb{R}^n$.

Airy equation   The equation $y'' - xy = 0$.

AKNS method   A procedure developed by Ablowitz, Kaup, Newell, and Segur (1973) that allows one, given a suitable scattering problem, to derive the nonlinear evolution equations solvable by the inverse scattering transform.

algebra   An *algebra* over a *field* F is a *ring* R which is also a finite dimensional *vector space* over F, satisfying $(ax)(by) = (ab)(xy)$ for all $a, b \in F$ and all $x, y \in R$.

algebraic equation   Let $f(x) = a_n x^n + a_{n-1} x^{n-1} + ... + a_1 x + a_0$ be a *polynomial* in $R[x]$, where R is a *commutative ring* with unity. The equation $a_n x^n + a_{n-1} x^{n-1} + ... + a_1 x + a_0 = 0$ is called an *algebraic equation*.

ALGOL   A programming language.

algorithm   A process consisting of a specific sequence of operations to solve certain types of problems.

alignment   In dealing with sequence data such as DNAs and proteins, one compares two such molecules by matching the sequences. Sequence *alignment* means finding optimal matching, defined by some criteria usually called "scores." Between two binary sequences, for example, the Hamming distance is a widely used score function.

almost complex manifold   A *manifold* with an *almost complex structure*.

almost complex structure   A *manifold* M is said to possess an *almost complex structure* if it carries a real differentiable tensor field J of type $(1, 1)$ satisfying $J^2 = -I$.

almost everywhere   A property holds *almost everywhere (a.e.)* if it holds everywhere except on a set of *measure zero*.

almost Hermitian   A *manifold* M with a Riemannian metric g invariant by the *almost complex structure* J, i.e., g and J satisfy

$$g(Ju, Jv) = g(u, v)$$

for any tangent vectors u and v.

almost Kähler   An *almost Kähler manifold* is an *almost Hermitian manifold* (M, J, g) such that the fundamental two-form Ω defined by $\Omega(u, v) = g(u, Jv)$ is *closed*.

almost periodic   A function $f(t)$ is called *almost periodic* if there exists $T(\epsilon)$ such that for any ϵ and every interval $I_\epsilon = (x, x + T(\epsilon))$, there is $x \in I_\epsilon$ such that, $|f(t + x) - f(t)| < \epsilon$.

α-limit set   Consider a *dynamical system* $u(t)$ in a *metric space* (M, d) which is described by a semigroup $S(t)$, i.e., $u(t) = S(t)u(0)$, $S(t + s) = S(t) \cdot S(s)$ and $S(0) = I$. The α-*limit set*, when it exists, of $u_0 \in M$, or $A \subset M$, is defined as

$$\alpha(u_0) = \bigcap_{s \le 0} \overline{\bigcup_{t \le s} S(-t)^{-1} u_0},$$

or

$$\alpha(A) = \bigcap_{s \le 0} \overline{\bigcup_{t \le s} S(-t)^{-1} A}.$$

Notice, $\phi \in \alpha(A)$ if and only if there exists a sequence ψ_n converging to ϕ in M and a sequence $t \to +\infty$, such that $\phi_n = S(t_n)\psi_n \in A$, for all n.

alphabet   A set of *letters* or other characters with which one or more languages are written.

alternating series   A *series* that alternates signs, i.e., of the form $\sum_n (-1)^n a_n$, $a_n \ge 0$.

alternation   For any *covariant tensor* field K on a *manifold* M the *alternation* A is defined as

$$(AK)(X_1, ..., X_r) = \frac{1}{r!} \sum_{\pi} (\text{sign } \pi)$$

$$K(X_{\pi(1)}, ..., X_{\pi(r)})$$

where the summation is taken over all $r!$ permutations π of $(1, 2, ..., r)$.

amplitude of a complex number   The angle θ is called the *amplitude of the complex number* $z = re^{i\theta} = r(\cos\theta + i\sin\theta)$.

amplitude of oscillation   The simplest equation of a linear oscillator is $m\frac{d^2x}{dt^2} = -kx$. It has the solution $x(t) = A\cos(t\sqrt{k/m} - c)$. A is called the *amplitude*.

analog   In contrast to digital, analog means a dynamic variable taking a continuum of values, e.g., a timepiece having hour and minute hands.

analog computation   Instead of using binary computation as in a digital computer, one uses a device which has continuous dynamic variables such as current (or voltage) in an electrical circuit, or displacement in a mechanical device.

analog computer   A device that computes using *analog computation*. A computer that operates with numbers represented by directly measurable quantities (e.g., voltage).

analog multiplier   Using analog computation to obtain the product, as output, of two (input) quantities.

analog variable   See *analog* and *analog computation*.

analytic dynamics   A *dynamical system* is called *analytic* if the coefficients of the *vector field* are *analytic functions*.

analytic function   A function $f(x)$ is called *analytic* at $x = a$ if it can be represented by a *power series*

$$f(x) = \sum_{n=0}^{\infty} c_n (x - a)^n \, ,$$

convergent for $|x| < r$, for some $r > 0$.

analytic geometry   A part of geometry in which algebraic methods are used to solve geometric problems.

analytic manifold   A *manifold M* such that its *coordinate transition functions* are *analytic*. See *chart*.

analytic structure   The set (atlas) of *coordinate* patches on an analytic *manifold M*. See *chart*.

analytical function   A function which relates the measure value $\widehat{C}_a$ to the instrument reading, X, with the value of all interferants, C_i, remaining constant. This function is expressed by the following regression of the calibration results:

$$\widehat{C}_a = f(X)$$

The *analytical function* is taken as equal to the inverse of the calibration function.

analytical index   The *analytical index* of an *elliptic* complex $\{D_p, E_p\}$ is defined as

$$index\{D_p, E_p\} = \sum_p (-1)^p dim \, ker \, \Delta_p,$$

where $\Delta_p = d\delta + \delta d$ is the *Laplacian* on p-forms.

analytical unit (analyser)   An assembly of subunits comprising: suitable apparatus permitting the introduction and removal of the gas, liquid, or solid to be analyzed and/or calibration materials; a measuring cell or other apparatus which, from the physical or chemical properties of the components of the material to be analyzed, gives signals allowing their identification and/or measurement; signal processing devices (amplification, recording) or, if need be, data processing devices.

angle   A system of two rays extending from the same point. The numerical measure of an angle is the degree (measured as a fraction of $360°$, the entire angle from a line to itself) or the radian ($= 180/\pi$ degrees).

angle between curves   Let $c_1(t)$ and $c_2(t)$ be two curves intersecting at t_0, i.e., $c_1(t_0) = c_2(t_0) = p$. The *angle between c_1 and c_2* at p is given by the *angle* between the two *tangent* vectors $\dot{c}_1(t_0)$ and $\dot{c}_2(t_0)$.

angle between lines   Let L_1 and L_2 be two nonvertical lines in the plane with *slopes m_1* and m_2, respectively. If θ, the angle from L_1 to L_2, is not a right angle, then

$$\tan \theta = \frac{m_2 - m_1}{1 + m_1 m_2} \, .$$

angle between planes   The *angle between two planes* is given by the *angle* between the two *normal vectors* to these planes.

angle between vectors   Let $\vec{u} \in \mathbb{R}^n$ and $\vec{v} \in \mathbb{R}^n$ be two vectors in $\mathbb{R}^n$. The *angle* θ between $\vec{u}$ and $\vec{v}$ is given by

$$\cos\theta = \frac{\vec{u} \cdot \vec{v}}{\|u\|\|v\|} ,$$

where the *dot product* $\vec{u} \cdot \vec{v} = \sum_{i=1}^{n} u_i v_i$ and the norm is $\|u\|^2 = \sum_{i=1}^{n} u_i^2$, if $\vec{u} = (u_1, ..., u_n)$ and $\vec{v} = (v_1, ..., v_n)$.

angle of depression   The *angle* between the horizontal plane and the line from the observer's eye to some object lower than the line of her eyes.

angle of elevation   The *angle* between the horizontal plane and the line from the observer's eye to some object above her eyes.

angle of incidence   The *angle* that a line (as a ray of light) falling on a surface or interface makes with the *normal vector* drawn at the point of incidence to that surface.

angle of reflection   The *angle* between a reflected ray and the *normal vector* drawn at the point of incidence to a reflecting surface.

angle of refraction   The *angle* between a refracted ray and the *normal vector* drawn at the point of incidence to the interface at which the refraction occurs.

angular   Measured by *angle*.

angular acceleration   The rate of change per unit time of *angular velocity*; i.e., if the *angular velocity* is represented by a *vector* $\vec{\omega}$ along the *axis of rotation*, then the *angular acceleration* $\vec{\alpha}$ is given by $\vec{\alpha} = \frac{d\vec{\omega}}{dt}$.

angular momentum, L (or moment of *momentum* of a particle about a point)   A *vector* quantity equal to the *vector product* of the position vector of the particle and its momentum, $L = r \times p$ where $r(t) = \frac{d}{dt} r(t)$ is the velocity vector and $p = m \cdot r$ is the momentum. For special *angular momenta* of particles in atomic and molecular physics different symbols are used.

angular variables   Let M be a *manifold* and S^1 the unit circle. A *smooth* map $\omega : M \to S^1$ is called an *angular variable* on M.

angular velocity   If a particle is moving in a plane, its *angular velocity* about a point in the plane is the *rate of change* per unit time of the *angle* between a fixed line and the line joining the moving particle to the fixed point.

anion   A monoatomic or polyatomic species having one or more elementary charges of the electron.

annihilation operator   For the *harmonic oscillator* with Hamiltonian $H = \frac{1}{2}(p^2 + \omega^2 q^2)$ the *annihilation operator* a is given by $a = \frac{1}{\sqrt{2\omega}}(p - i\omega q)$. The creation operator a^* is given by $a^* = \frac{1}{\sqrt{2\omega}}(p + mi\omega q)$. Then $H = \frac{\omega}{2}(aa^* + a^*a)$ and we have $a\psi(N) = \sqrt{N}\psi(N-1)$ and $a^*\psi(N) = \sqrt{N+1}\psi(N+1)$.

annihilator   Let X be a *vector space*, X^* its *dual vector space*, and Y a *subspace* of X. The *annihilator* $M^\perp$ of M is defined as $M^\perp = \{f \in X^* \mid f(x) = 0, \text{ for all } x \in M\}$.

annulus   The region of a plane bounded by two concentric circles in the plane. Let $R > r$, the *annulus* **A** determined by the two circles of radius R and r, respectively, (centered at 0) is given by

$$\mathbf{A} = \{\vec{x} = (x, y) \in \mathbb{R}^2 \mid r < \|\vec{x}\| < R\}$$

where $\|\vec{x}\| = \sqrt{x^2 + y^2}$.

anomalies   In *quantum field theories anomalies* are quantum effects of conservation laws; i.e., if one has a conservation law at the classical level which is not true at the quantum level, this is expressed by an anomaly, e.g., scale invariance is violated when quantized, which gives rise to a scale factor, the anomaly.

Anosov system   A *diffeomorphism* on a *manifold* which has a *hyperbolic* structure everywhere is called an *Anosov system*.

ansatz   An "assumed form" for a solution; a simplified assumption.

antibody   A protein (*immunoglobulin*) produced by the immune system of an organism in response to exposure to a foreign molecule (*antigen*) and characterized by its specific binding to a site of that molecule (antigenic determinant or *epitope*).

anticommutator   If A, B are two linear *operators*, their *anticommutator* is $\{A, B\} = AB + BA$.

antiderivation   A linear operator T on a graded algebra $(A, \cdot)$ satisfying $T(a \cdot b) = Ta \cdot b + (-1)^{(\text{degree of } b)} a \cdot Tb$ for all $a, b \in A$.

antiderivative   A function $F(x)$ is called an *antiderivative* of a function $f(x)$ if $F'(x) = f(x)$.

antigen   A substance that stimulates the immune system to produce a set of specific *antibodies* and that combines with the *antibody* through a specific binding site or *epitope*.

antimatter   Matter composed of *antiparticles*.

antiparticle   A subatomic particle identical to another subatomic particle in mass but opposite to it in the electric and magnetic properties.

antiselfdual   A gauge field F such that $F = - * F$, where $*$ is the Hodge-star operator.

aphelion   The point in the path of a celestial body (as a planet) that is farthest from the sun.

apogee   The point in the orbit of an object (as a satellite) orbiting the earth that is the greatest distance from the center of the earth.

applied potential   The difference of potential measured between identical metallic leads to two electrodes of a cell. The applied potential is divided into two *electrode potentials*, each of which is the difference of potential existing between the bulk of the solution and the interior of the conducting material of the electrode, an $i R$ or ohmic potential drop through the solution, and another ohmic potential drop through each electrode.

In the electroanalytical literature this quantity has often been denoted by the term *voltage*, whose continued use is not recommended.

approximate solution   Consider the *differential equation* (*) $x' = f(x, t)$, $x \in \Omega \subset \mathbb{R}^n, t \in [a, b]$. The *vector* valued function $y(t)$ is an ϵ-*approximate solution* of (*) if $\|y'(t) - f(t, y(t))\|_{\mathbb{R}^n} < \epsilon$, for all $t \in [a, b]$.

arc   (**1**) A segment, or piece, of a curve.
(**2**) The image of a closed interval $[a, b]$ under a one-to-one, *continuous* map.

arc length   Let $\sigma : [a, b] \to \mathbb{R}^n$ be a C^1 curve. The *arc length* $l(\sigma)$ of σ is defined as

$$l(\sigma) = \int_a^b \|\sigma'(t)\| dt.$$

arccosecant   The inverse trigonometric function of *cosecant*. The *arccosecant* of a number x is a number y whose *cosecant* is x, written as $y = \csc^{-1}(x) = \text{arc } \csc(x)$, i.e., $x = \csc(y)$.

arccosine   The inverse trigonometric function of *cosine*. The *arccosine* of a number x is a number y whose *cosine* is x, written as $y = \cos^{-1}(x) = \text{arc } \cos(x)$, i.e., $x = \cos(y)$.

arccotangent   The inverse trigonometric function of *cotangent*. The *arccotangent* of a number x is a number y whose *cotangent* is x, written as $y = \cot^{-1}(x) = \text{ctn}^{-1}(x) = \text{arc } \cot(x)$, i.e., $x = \cot(y)$.

arcsecant   The inverse trigonometric function of *secant*. The *arccosecant* of a number x is a number y whose *secant* is x, written as $y = \sec^{-1}(x) = \text{arc } \sec(x)$, i.e., $x = \sec(y)$.

arcsine   The inverse trigonometric function of *sine*. The *arcsine* of a number x is a number y whose *sine* is x, written as $y = \sin^{-1}(x) = \text{arc } \sin(x)$, i.e., $x = \sin(y)$.

arctangent   The inverse trigonometric function of *tangent*. The *arctangent* of a number x is a number y whose *tangent* is x, written as $y = \tan^{-1}(x) = \text{arc } \tan(x)$, i.e., $x = \tan(y)$.

area of surface   Consider the surface S given by $z = f(x, y)$ that projects onto the bounded region D in the xy-plane. The area $A(S)$ of the surface S is given by

$$A(S) = \iint_D \sqrt{f_x^2 + f_y^2 + 1} \, dD.$$

area under curve   Let a curve be given by $y = f(x)$, $a \le x \le b$. The *area A under this curve* from a to b is given by the integral

$$A = \int_a^b f(x)dx \,.$$

Argand diagram   The basic idea of *complex numbers* is credited to Jean Robert Argand, a Swiss mathematician (1768–1822). An *Argand diagram* is a rectangular coordinate system in which the *complex number* $x + iy$ is represented by the point whose coordinates are x and y. The x-axis is called real axis and the y-axis is called imaginary axis.

argument   The collection of elements satisfying some relation r is called the set of *arguments* of r.

argument of complex number   See *amplitude of a complex number.*

arithmetic   The study of the positive integers $1, 2, 3, 4, 5, \ldots$ under the operations of addition, subtraction, multiplication, and division.

arithmetic difference   The *arithmetic difference* of two numbers a and b is $|a - b|$.

arithmetic division   To determine the *arithmetic quotient* $\left[\frac{a}{b}\right]$ of two nonnegative integers a and b, where $[x]$ is the greatest integer, which is not bigger than x.

arithmetic mean   The *arithmetic mean* of n numbers $a_1, a_2, \ldots, a_n$ is

$$x = \frac{a_1 + a_2 + \cdots + a_n}{n} \,.$$

arithmetic progression   A *sequence* of numbers $a_1, a_2, \ldots, a_n, \ldots$ in which each following number is obtained from the preceding number by adding a given number r, i.e., $a_n = a_1 + (n - 1)r$.

arithmetic quotient   See *arithmetic division.*

arithmetic sequence   A sequence $a, (a + d), (a + 2d), \cdots, (a + nd), \cdots$, in which each term is the arithmetic mean of its neighbors.

arithmetic sum   The sum of an *arithmetic sequence* $\sum_{n=0}^{N}(a + nd)$.

arity   The number of arguments of a relation.

array   A display of objects in some regular arrangements, as a rectangular array or *matrix* in which numbers are displayed in rows and columns, or an arrangement of statistical data in order of increasing (or decreasing) magnitude.

array index   In a rectangular array such as a *matrix* the element in the ith row and jth column is indexed as a_{ij}.

artificial intelligence   A branch of computer science dealing with the simulation of intelligent behavior of computers.

ascending sequence   A *sequence* $\{a_n\}$ is called *ascending* (increasing) if each term is greater than the previous term, i.e., $a_n \ge a_{n-1}$. If $a_n > a_{n-1}$ then it is called *monotone ascending/increasing*.

ASCII   American Standard Code for Information Interchange. A code for representing alphanumeric information.

Ascoli   Giulio Ascoli (1843–1896), Italian analyst.

Ascoli's theorem   Let $\{f_n\}$ be a family of *uniformly bounded* equicontinuous functions on $[0, 1]$. Then some subsequence $\{f_{n(i)}\}$ converges uniformly on $[0, 1]$.

assembler   A computer program that automatically converts instructions written in assembly language into computer language.

assembly   Computation of a finite element *stiffness matrix* A from the *element matrices* A_K belonging to the cells K of the underlying *mesh* Ω_h. The general formula is

$$A = \sum_{K \in \Omega_h} I_K A_K I_K^T,$$

where the I_K are rectangular matrices reflecting the association of local and global degrees of freedom.

assembly language   A programming language that consists of instructions that are mnemonic codes for corresponding machine language instructions.

associated bundle   If $(P, M, p : G)$ is a *principal bundle* and $\lambda : G \times F \to F$ is a *left action* of the *Lie group* G and a *manifold F*, then one can define a right action of G on $P \times F$ using the canonical right action R of G on P as follows:

$$(p, f) \cdot g = (R_g p, \lambda(g^{-1}, f)).$$

The quotient space $P \times_\lambda F \equiv (P \times F) \backslash G$ has a canonical structure of a bundle $(P \times_\lambda F, M, \pi; F)$ and it is called an *associated bundle to P*.

Trivializations of P induce trivializations of $P \times_\lambda F$, *principal connections* on P induce connections on $P \times_\lambda F$, right invariant *vector fields* on P induce vector fields on $P \times_\lambda F$. The transition functions of $P \times_\lambda F$ are the same transition functions of P represented on F by means of the action λ.

association   The assembling of separate *molecular entities* into any aggregate, especially of oppositely charged free ions into *ion pairs* or larger and not necessarily well-defined clusters of ions held together by electrostatic attraction. The term signifies the reverse of *dissociation*, but is not commonly used for the formation of definite *adducts* by *colligation* or *coordination*.

associative   Describing an operation among objects $x, y, z, ...$, denoted by $\bullet$, such that $(x \bullet y) \bullet z = x \bullet (y \bullet z)$. For example, addition and multiplication of numbers *associative* $(x + y) + z = x + (y + z)$, $(x \cdot y) \cdot z = x \cdot (y \cdot z)$, for all numbers x, y, z.

asymptote   A straight line associated with a plane curve such that as a point moves along an infinite branch of the curve the distance from the point to the line approaches zero and the *slope* of the tangent to the curve at the point approaches the *slope* of the line.

asymptote to the hyperbola   The standard form of the equation of the *hyperbola* in the plane is $x^2/a^2 - y^2/b^2 = 1$. The lines $y = bx/a$ and $y = -bx/a$ are its *asymptotes*.

asymptotic freedom   Property of *quantum field* theories which says that interactions are weak at high energies (momenta). Mathematically this means that suitably normalized correlation functions tend to the correlation function of a free theory when the momenta go to infinity.

asymptotic series   A function $f(x)$ on $(0, a)$, $a > 0$, is said to have $\sum a_n x^n$ as *asymptotic series (expansion)* as $x \downarrow 0$, written $f(x) \sim \sum a_n x^n$, $x \downarrow 0$, if, for each N

$$\lim_{x \downarrow 0} \left[f(x) - \sum_{n=0}^{N} a_n x^n \right] \bigg/ x^N = 0.$$

asymptotically dense   Let $\{V_h\}_{h \in \mathbb{H}}$, $\mathbb{H}$ some index set, be a family of finite dimesional subspaces of the *Banach space V*. This family is called *asymptotically dense*, if

$$\overline{\bigcup_{k \in \mathbb{H}} V_h} = V,$$

where the closure is with respect to the norm of V.

asymptotically equal   Two functions f and g are said to be *asymptotically equal* (at infinity) if, for every $N > 0$, we have

$$\lim_{x \to \infty} x^N (f(x) - g(x)) = 0.$$

asymptotically optimal   A finite-element solution of a variational problem is regarded as *asymptotically optimal*, if it is *quasi optimal*, provided that the *meshwidths* of the underlying triangulations stay below a certain threshold.

asymptotically stable   Let X be a *vector field* on the *manifold M* and F_t its flow. A *critical point* m_0 of X is called *asymptotically stable* if there is a *neighborhood V* of m_0 such that for each $m \in V$ there exists an *integral curve* $c_m(t)$ of X starting at m for all $t > 0$, $F_t(V) \subset F_s(V)$ if $t > s$ and $\lim_{t \to +\infty} F_t(V) = \{m_0\}$. m_0 is asymptotically unstable if it is *asymptotically stable* as $t \to -\infty$.

Atiyah   Sir Michael Atiyah (1929–), Differential Geometer/Mathematical Physicist, Professor emeritus University of Edinburgh. Fields Medal 1966, Knighted 1983.

Atiyah-Singer index theorem   The *Atiyah-Singer index theorem* gives the equality between the *analytic index* and the topological index of an elliptic complex over a compact *manifold*. The *analytical index* of an elliptic complex $\{D_p, E_p\}$ is defined as

$$\text{index}\{D_p, E_p\} = \sum_p (-1)^p \dim \ker \Delta_p.$$

The topological index of an elliptic complex $\{D_p, E_p\}$ is defined as

$$\text{topindex}\{D_p, E_p\}$$
$$= \int_{\psi(M)} ch(\Sigma(D)) \Lambda \rho^* tod M,$$

where $ch(\Sigma(D))$ is the Chern character of the symbol bundle $\Sigma(D)$, ρ the projection of the compactified cotangent bundle $\psi(M)$ onto M, and $tod(M)$ the Todd class.

Theorem (Atiyah-Singer) $\text{index}\{D_p, E_p\} = \text{topindex}\{D_p, E_p\}$.

atlas   An *atlas* on a *manifold M* is a collection of *charts* whose domains cover M.

atmosphere (of the earth)   The entire mass of air surrounding the earth which is composed largely of nitrogen, oxygen, water vapor, clouds (liquid or solid water), carbon dioxide, together with trace gases and aerosols.

atomic formula   A term $f(t_1, \ldots, t_n)$, where f is a relation.

Comment: Note this is "atomic" in the computer science and linguistic, not the chemistry, senses.

atomic units   System of units based on four base quantities: length, mass, charge, and action (angular momentum) and the corresponding base units the Bohr radius, a_0, rest mass of the electron, m_e, elementary charge, e, and the Planck constant divided by 2π, $\hbar$.

attenuance, D   Analogous to *absorbance*, but taking into account also the effects due to scattering and luminescence. It was formerly called extinction.

attractor   Consider a dynamical system $u(t)$ in a metric space (M, d) which is described by a semigroup $S(t)$, i.e., $u(t) = S(t)u(0)$, $S(t + s) = S(t) \cdot S(s)$ and $S(0) = I$. An *attractor* for $u(t)$ is a set $A \subset M$ with the following properties:

(i.) A is an invariant set, i.e., $S(t)A = A$, for all $t \geq 0$.

(ii.) A possesses an *open neighborhood U* such that, for every $u_0 \in U$, $S(t)u_0$ converges to A as $t \to \infty$:

$$dist(S(t)u_0, A) \to 0 \ as \ t \to \infty,$$

where the distance $d(x, A) = \inf_{y \in A} d(x, y)$.

augmented matrix   If a system of linear equations is written in *matrix* form $A\vec{x} = \vec{b}$, then the *matrix* $[A|\vec{b}]$ is called the *augmented matrix*.

autocatalysis   A reaction in which the product also serves as a *catalyst*. Hence this reaction is nonlinear with a positive feedback. *Autocatalysis* is an important ingredient for an oscillatory chemical reaction.

automorphism   An *isomorphism* of a set with itself. Also an *isomorphism* of an object of a *category* into itself.

autonomous system   A system of *differential equations* $\frac{d\vec{x}}{dt} = F(\vec{x})$ is called *autonomous* if the independent variable t does not appear explicitly in the function F.

autoparallel   A *vector field X* on a *Riemannian manifold M* is called *autoparallel* along a curve $c(t)$ if the *covariant derivative* of X along c vanishes, i.e., $\nabla_{\dot{c}(t)} X = 0$. In local coordinates

$$\ddot{c}^i(t) + \Gamma^i_{jk}(c(t))\dot{c}^j(t)\dot{c}^k(t) = 0 \, .$$

Hence $\dot{c}$ is autoparallel along c if c is a *geodesic*.

average   The *average* value of a *function* $f(x)$ over an interval $[a, b]$ is given by the number

$$\frac{1}{b-a} \int_a^b f(x)\, dx$$

axial gauge   For a fixed *vector n*, there exists a *gauge transformation* $A \rightarrow A'$ such that $n \cdot A'(x) = 0$. This is called the *axial gauge*.

axiom   A statement that is accepted without proof. The *axioms* of a mathematical theory are the basic propositions from which all other propositions can be derived.

axis of rotation   A straight line about which a body or geometric figure is rotated.

axis of symmetry   A straight line with respect to which a body or geometric figure is symmetrical.

azimuth   Horizontal direction expressed as the angular distance between the direction of a fixed point and the direction of an object.

  In *polar coordinates* (r, θ) in the plane the polar angle θ is called the *azimuth* of the point P.

B

Bäcklund transformations Transformations between solutions of *differential equations*, in particular *soliton* equations. They can be used to construct nontrivial solutions from the trivial solution.

Formally: Two *evolution equations* $u_t = K(x, u, u_1, ..., u_m)$ and $v_t = G(y, v, v_1, ..., v_n)$ are said to be *equivalent under a Bäcklund transformation* if there exists a transformation of the form $y = \psi(x, u, u_1, ..., u_n)$, $v = \phi(x, u, u_1, ..., u_n)$.

bag An unordered collection of elements, including duplicates, each of which satisfies some property. An enumerated bag is delimited by braces ($\{x\}$).

Comment: Unfortunately, the same delimiters are used for sets and bags. See also *list, sequence, set,* and *tuple.*

Baire space A space which is not a countable union of nowhere dense subsets. Example: A complete metric space is a Baire space.

Baker-Campell-Hausdorff formula For any $n \times n$ *matrices* A, B we have

$$e^{-sA}Be^{sA} = B + s[A, B] + \frac{s^2}{2}[A, [A, B]] + \cdots$$

balanced set A subset M of a vector space V over $\mathbb{R}$ or $\mathbb{C}$ such that $\alpha x \in M$, whenever $x \in M$ and $|\alpha| \leq 1$.

ball Let (X, d) be a *metric space*. An *open ball* $B_a(x_0)$ of radius a about x_0 is the set of all $x \in X$ such that $d(x, x_0) < a$. The *closed ball* $\bar{B}_a(x_0) = \{x \in X \mid d(x, x_0) \leq a\}$.

Banach Stefan Banach (1892–1945). Polish algebraist, analyst, and topologist.

Banach algebra A *Banach space X* together with an internal operation, usually called multiplication, satisfying the following: for all x, $y, z \in X, \alpha \in \mathbb{C}$

(i.) $x(yz) = (xy)z$

(ii.) $(x + y)z = xz + yz$, $x(y+z) = xy + xz$

(iii.) $\alpha(xy) = (\alpha x)y = x(\alpha y)$

(iv.) $\|xy\| \leq \|x\|\|y\|$

(v.) X contains a *unit* element $e \in X$, such that $xe = ex = x$

(vi.) $\|e\| = 1$.

Banach fixed point theorem Let (X, d) be a *complete metric space* and $T : X \to X$ a *contraction* map. Then T has a unique fixed point $x_0 \in X$, i.e., $T(x_0) = x_0$.

Banach manifold A *manifold* modeled on a *Banach space.*

Banach space A *normed vector space* which is *complete* in the *metric* defined by its norm.

barrel A subset of a topological vector space which is *absorbing, balanced, convex,* and *closed.*

barreled space A topological *vector* space E is called *barreled* if each *barrel* in E is a *neighborhood* of $0 \in E$, i.e., the barrels form a *neighborhood* base at 0.

barycenter The *barycenter* (center of mass) of the *simplex* $\sigma = (a_0, ..., a_p)$ is the point

$$b_\sigma = \frac{1}{p+1}(a_0 + \cdots + a_p).$$

barycentric coordinates Let $p_0, ..., p_n$ be $n + 1$ points in n-dimensional Euclidean space E^n that are not in the same hyperplane. Then for each point $x \in E^n$ there is exactly one set of real numbers $(\lambda_0, ..., \lambda_n)$ such that

$$x = \lambda_0 p_0 + \lambda_1 p_1 + \cdots + \lambda_n p_n$$

and

$$\lambda_0 + \lambda_1 + \cdots + \lambda_n = 1.$$

The numbers $(\lambda_0, ..., \lambda_n)$ are called *barycentric coordinates* of the point x.

base for a topology   A collection B of *open* sets of a *topological space* T is a *base for the topology* of T if each open set of T is the union of some members of B.

base space   Let $\pi : E \to B$ be a smooth *fiber bundle*. The *manifold* B is called the *base space* of π.

basis graph   A subgraph $\mathsf{G}'(\mathcal{V}, \mathcal{E}')$ of $\mathsf{G}(\mathcal{V}, \mathcal{E})$ such that $\mathcal{E}' \subset \mathcal{E}$, and that all pairs of nodes $\{v_i, v_j\} \subset \mathcal{V}$ in G and G' are connected (i, j indices).

basis, Hamel   A maximal *linear* independent subset of a *vector* space X. Such a basis always exists by *Zorn's lemma*.

basis of a vector space   A subset E of a *vector space* V is called a *basis* of V if each *vector* $x \in V$ can be uniquely written in the form

$$x = \sum_{i=1}^{n} a_i e_i , \quad e_i \in E.$$

The numbers $a_1, ..., a_n$ are called *coordinates* of the *vector* x with respect to the basis E.
Example: $E = (e_1, ..., e_n)$ with $e_1 = (1, 0, ..., 0), e_2 = (0, 1, 0, ..., 0), ..., e_n = (0, 0, ..., 1)$ is the *standard* basis of $V = \mathbb{R}^n$.

Bayes formula   Suppose A and $B_1, ..., B_n$ are events for which the probability $P(A)$ is not 0, $\sum_{i=1}^{n} P(B_i) = 1$, and $P(B \text{ and } B_j) = 0$ if $i \neq j$. Then the conditional probability $P(B_j|A)$ of B_j given that A has occurred is given by

$$P(B_j|A) = \frac{P(B_j)P(A|B_j)}{\sum_{i=1}^{n} P(B_i)P(A|B_i)} .$$

beam equation $u_t + u_{xxxx} = 0.$

Becchi-Rouet-Stora-Tyutin (BRST) transformation   In nonabelian gauge theories the effective action functional is no longer *gauge invariant*, but it is invariant under the BRST transformation $\mathbf{s}$

$$\mathbf{s}A = d\eta + [A, \eta],$$

$$\mathbf{s}\eta = -\frac{1}{2}[\eta, \eta] , \quad \mathbf{s}\bar{\eta} = b , \quad \mathbf{s}b = 0.$$

where A is the *vector* potential and η and $\bar{\eta}$ are the ghost and anti-ghost fields, respectively. One of the main properties of the BRST transformation $\mathbf{s}$ is its nilpotency, $\mathbf{s}^2 = 0$.

Belousov-Zhabotinskii reaction   A chemical reaction which involves the oxidation of malonic acid by bromate ions, BrO_3^-, and catalyzed by cerium ions, which has two states Ce^{3+} and Ce^{4+}. With appropriate dyes, the reaction can be monitored from the color of the solution in a test tube. This is the first reaction known to exhibit sustained chemical oscillation. Spatial pattern has also been observed in BZ reaction when diffusion coefficients for various species are in the appropriate region.

Benjamin-Ono equation   The *evolution equation*

$$u_t = Hu_{xx} + 2uu_x$$

where H is the *Hilbert transform*

$$(Hf)(x) = \frac{1}{\pi} \int_{-\infty}^{+\infty} \frac{f(\xi)}{\xi - x} d\xi .$$

Berezin integral   An integration technique for Fermionic fields in terms of anticommuting algebras. Let $B = B_+ \oplus B_-$ be a DeWitt algebra (super algebra) and $y \mapsto f(y)$ a supersmooth function from B_- into B. Then $f(y) = f_0 + f_1 y$ with f_0, f_1 in B. The *Berezin integral* of f on B_- is

$$\int_{B_-} f(y)dy = c_1 f_1$$

where c is a constant independent of f.

Bergman kernel   Let M be an n-dimensional *complex manifold* and H the *Hilbert space* of holomorphic n-forms on M. Let $h_0, h_1, h_2, ...$ be a *complete orthonormal basis* of H and $z^1, ..., z^n$ a local coordinate system of M. The *Bergman kernel form* K is defined by

$$K = K^* dz^1 \wedge \cdots \wedge dz^n \wedge d\bar{z}^1 \wedge \cdots \wedge d\bar{z}^n,$$

the function K^* is the *Bergman kernel function* on M.

Bergman metric   Let M be an n-dimensional *complex manifold*, in any complex coordinate system $z^1, ..., z^n$. The Kähler metric

$$ds^2 = 2 \sum g_{\alpha\bar{\beta}} dz^\alpha d\bar{z}^\beta .$$

with

$$g_{\alpha\bar{\beta}} = \partial^2 \log K^* / \partial z^\alpha \partial \bar{z}^\beta$$

is called the *Bergman metric* of M.

Bernoulli equation   Let $f(x), g(x)$ be continuous functions and $n \neq 0$ or 1, the *Bernoulli equation* is $\frac{dy}{dx} + f(x)y + g(x)y^n = 0$.

Bernoulli numbers   The coefficients of the *Bernoulli polynomials*.

Bernoulli polynomials

$$B_m(z) = \sum_{k=0}^{m} \binom{m}{k} B_k z^{m-k}.$$

The coefficients B_k are called *Bernoulli numbers*. The *Bernoulli polynomials* are solutions of the equations

$$u(z+1) - u(z) = mz^{m-1}, \quad m = 2, 3, 4, \ldots$$

Bessel equation   The *differential equation*

$$z^2 \frac{d^2 y}{dz^2} + z \frac{dy}{dz} + (z^2 - n^2)y = 0.$$

Bessel function   For $n \in \mathbb{Z}$ the nth *Bessel function* $J_n(z)$ is the coefficient of t^n in the expansions $e^{z[t-1/t]/2}$ in powers of t and $1/t$. In general,

$$J_n(z) = \frac{1}{\pi} \int_0^{\pi} \cos(nt - z \sin t) dt$$

$$= \sum_{r=0}^{\infty} \frac{(-1)^r}{r! \Gamma(n+r+1)} \left(\frac{z}{2}\right)^{n+2r}.$$

$J_n(z)$ is a solution of the *Bessel equation*.

beta function   The *beta function* is defined by the *Euler integral*

$$B(z, y) = \int_0^1 t^{z-1}(1-t)^{y-1} dt$$

and is the solution to the *differential equation*

$$-(z+y)u(z+1) + zu(z) = 0.$$

The beta function satisfies

$$B(z, y) = \frac{\Gamma(z) \cdot \Gamma(y)}{\Gamma(z+y)}.$$

See *gamma function*.

Betti number   Let H_p be the pth *homology group* of a simplicial complex K. H_p is a finite dimensional *vector* space and the dimension of H_p is called the pth *Betti number* of K.

Let M be a *manifold* and $H^p(M)$ the pth *De Rham cohomology group*. The dimension of the finite dimensional *vector* space $H^p(M)$ is called the pth *Betti number* of M.

Bianchi's identities   In a *principal* fiber bundle $P(M, G)$ with connection 1-form ω and curvature 2-form $\Omega = D\omega$ (D is the *exterior covariant derivative*), Bianchi's identity is $D\Omega = 0$.

In terms of the *scalar* curvature R on a *Riemannian manifold*, Bianchi's identity is $R(X, Y, Z) + R(Z, X, Y) + R(Y, Z, X) = 0$.

bifurcation   The qualitative change of a *dynamical system* depending on a control parameter.

bifurcation point   Let X_λ be a *vector field* (dynamical system) depending on a parameter $\lambda \in \mathbb{R}^n$. As λ changes the dynamical system changes, and if a qualitative change occurs at $\lambda = \lambda_0$, then λ_0 is called a *bifurcation point* of X_λ.

bi-Hamiltonian   A *vector field* X_H is called *bi-Hamiltonian* if it is *Hamiltonian* for two independent symplectic structures ω_1, ω_2, i.e., $X_H(F) = \omega_1(H, F) = \omega_2(H, F)$ for any function F.

bijection   A map $\phi : A \to B$ which is at the same time *injective* and *surjective*.

A map ϕ is *invertible* if and only if it is a bijection.

bijective   A function is *bijective* if it is both injective and surjective, i.e., both one-to-one and *onto*. See also *onto*, *into*, *injective*, and *surjective*.

bilateral network   There are two classes of simple neural networks, the feedforward and feedback (forming a loop) networks. In both cases, the connection between two connected units is unidirectional. In a bilateral network, the connection between two connected units is bi-directional.

bilinear map   Let X, Y, Z be *vector spaces*. A map $B : X \times Y \to Z$ is called *bilinear* if it is *linear* in each factor, i.e.,

$$B(\alpha x + \beta y, z) = \alpha B(x, z) + \beta B(y, z),$$

$$B(z, \alpha x + \beta y) = \alpha B(z, x) + \beta B(z, y).$$

binary   A *binary number system* is based on the number 2 instead of 10. Only the digits 0 and 1 are needed. For example, the binary number $101110 = 1 \cdot 2^5 + 0 \cdot 2^4 + 1 \cdot 2^3 + 1 \cdot 2^2 + 0 \cdot 2^0 = 46$ in decimal notation.

A *binary operation* is an operation that depends on two objects. Addition and, multiplication are binary operations.

binomial   The formal sum of two terms, e.g., $x + y$.

binomial coefficients   The coefficients in the expansion of $(x + y)^n$. The $(k + 1)$st binomial coefficient of order n is the coefficient of $x^{n-k} y^k$, and it is given by

$$\binom{n}{k} = \frac{n!}{k!(n-k)!}.$$

This is also the number of combinations of n things k at a time.

bioassay   A test to determine whether a chemical has any biological function (sometimes also called activity). This is usually accomplished by a set of chemical reactions leading to an observable change in biological systems or in test tubes.

biochemical graph   A set of *biochemical reactions*, their participating molecules, and labels for reactions, molecules, and subgraphs, represented as a *graph*.

Comment: The considered biochemical graphs are sometimes hypergraphs, mathematically. However, the key results and *algorithms* of the two objects are equally applicable; the common usage in computer science is to use the word "graph." Notice this is simply the biochemical network with an empty parameter set. See also *biochemical network*.

biochemical motif   A *motif* describing a biochemical relationship between two compounds in the donor-acceptor formalism.

Comment: The constraint for bimolecular relationship permits use of the common donor-acceptor language. A reaction may have more than one such relationship. Note that the biochemical donor-acceptor relationship is often opposite to that of the chemical one: thus a phosphoryl donor is a nucleophile acceptor. See also *chemical, dynamical, functional, kinetic, mechanistic, phylogenetic, regulatory, thermodynamic,* and *topological motifs.*

biochemical network   A mathematical network $\mathsf{N}(\mathcal{V}, \mathcal{E}, \mathcal{P}, \mathcal{L})$ representing a system R of *biochemical reactions*, their participating molecular species; descriptive, transformational, thermodynamic, kinetic, and dynamic parameters describing the *reactions* singly and composed together; and labels giving the names of *reactions*, molecules, and subnetworks. $\mathcal{V}$ is the bipartite set of vertices: $\mathcal{V}_m$ representing molecular species; $\mathcal{V}_r$ representing reactive conjunctions of molecules, $\mathcal{V} = \mathcal{V}_m \cup \mathcal{V}_r$. $\mathcal{E} = \mathcal{E}_s \cup \mathcal{E}_d \cup \mathcal{E}_c$ is the set of relations between molecule and reactive conjunction vertices, $e(\lambda, v_{m,i}, v_{r,j}) \in \mathcal{E}$, where for each pair $(v_{m,i}, v_{r,j})$, λ is one and only one of $\{s, d, c\} = \Lambda$: a molecule is a member of the set of coreacting species that appear sinistralaterally, dextralaterally, or catalytically in the reaction equation. Members of the parameter set $\mathcal{P}$ apply to vertices, edges, and connected graphs of vertices and edges as biochemically appropriate and as such information is available. If there are no parameters ($\mathcal{P} = \emptyset$), the network $\mathsf{N}(\mathcal{V}, \mathcal{E}, \mathcal{P}, \mathcal{L})$ reduces to its graph $\mathsf{N}'(\mathcal{V}, \mathcal{E}, \mathcal{L})$. Labels apply to vertices, edges, and subnetworks and take the form of one of the elements of $\{l_{m,i}, l_{r,j}, l_{((m,i),(r,j))}, l_{\{\mathcal{V}_m, \mathcal{V}_r, \mathcal{E}\}}\}$.

Comment: The network is a *biochemical graph* whose nodes, edges, and subgraphs have qualitative and quantitative parameters. Thus concentration is a property of a compound node; $\Delta G^{0'}$ is a property of a set of compound and reactive conjunction nodes, and their incident edges; k_{cat} is a property of the edge joining an enzyme to its reaction; molecular structure is a property of a compound node; etc. Not all nodes or edges need be so marked; and in fact much known information is at present unavailable electronically.

biochemical reaction   A *biochemical reaction* is any spontaneous or catalyzed transformation of covalent or noncovalent molecular bonds which occur in biological systems, written as a balanced, formal reaction equation, including all participating molecular species, whose kinetic order equals the sum of the partial orders of the reactants, including the active form of any catalyst(s). Equally, a composition of a set of bichemical reactions.

Comment: The definition places no restrictions on the level of resolution of the description, or size and complexity of reacting species; thus, it permits the recursive specification of processes. "Distinct origin" means molecular species arising from different precursors. Thus, two protons, if one came from water and the other from a protein, would be individually recorded in the equation. By "kinetic significance" is meant any molecular species which at any concentration contributes a term to the empirical rate law of the overall reaction. From the empirical rate law, the reaction's apparent kinetic order is the sum of the partial orders of the reactants (including catalysts). The restriction to active forms of the catalyst includes those instances where the catalyst must be activated, by either covalent modification or ligand binding, or is inhibited by those means, so that not all molecules present are equally capable of catalysis. The definition places no restrictions on the level of resolution of the description, or size and complexity of reacting species, thus permitting the recursive specification of processes. The recursion scales over any size or complexity of process.

bioinformatics   See *computational biology.*

biological functions   The roles a molecule plays in an organism.

Comment: By *function* (called here *biological function* to distinguish it from the mathematical sense of *function*), biologists mean both how a molecule interacts with its milieu and what results from those interactions. The results are often decomposed into biochemical, physiological, or genetic functions, but it is equally plausible to consider, for example, the ultrastructural function of a molecule (what part of the cell's microanatomy does it build, how strong is it, etc.). What is critical to realize is that a molecule always has more than one function; at a minimum it must be made and degraded.

biometrics   The field of study that uses mathematical and statistical tools to solve biological problems and solving mathematical and statistical problems arising from biology. In recent years, it has a much narrower meaning in practice: it mainly deals with statistical analysis and methodology applicable to biology and medicine.

bipartite   Describing a graph $G(\mathcal{V}, \mathcal{E})$ whose *chromatic number* $\chi(G) = 2$.

Comment: Informally, a graph will be bipartite if it has two distinct sets of nodes and if nodes of each type are always adjacent to the other. Thus the *biochemical graph* is *bipartite* because it has a set of compound nodes and a set of reactive conjunction nodes, and each is connected to the other.

black hole (general relativity)   A hypothetical object in space with so intense a gravitational field that light and matter cannot escape.

blob   See *denser subgraph.*

block   A portion of a *macromolecule*, comprising many *constitutional units* that has at least one feature which is not present in the adjacent portions. Where appropriate, definitions relating to *macromolecule* may also be applied to *block.*

block matrix   If a *matrix* is partitioned in submatrices it is called a *block matrix.*

Bogomolny equations   The self-dual Yang-Mills-Higgs equations are called *Bogomolny equations.* They are

$$F_{\mu\nu} = \frac{1}{2}\epsilon_{\mu\nu\rho\sigma} F_{\rho\sigma}$$

where $F_{ab} = \frac{1}{2}\epsilon_{abc}B_c$, $F_{a4} = D_a\phi$, $a, b, c = 1, 2, 3$. The solutions of the *Bogomolny equations* are called magnetic monopoles.

Bohr radius $r_B = \frac{h^2}{4\pi^2 m_e e^2} = 0.529 \times 10^{-8}$ cm, where h is the Planck constant, m_e is the rest mass of the electron, and e is the electron charge.

Boltzmann constant The fundamental physical constant $k = R/L = 1.380 = 658 \times 10^{-23} \text{JK}^{-1}$, where R is the gas constant and L the Avogadro constant. In the ideal gas law $PV = NkT$, where P is the pressure, V the volume, T the *absolute temperature*, N the number of *moles*, and k is the *Boltzmann constant*.

Boltzmann equation *Boltzmann's equation* for a density function $f(x, v, t)$ is the equation of continuity (mass conservation)

$$\frac{\partial f}{\partial t}(x, v, t) + \dot{x}\frac{\partial f}{\partial x}(x, v, t) + \dot{v}\frac{\partial f}{\partial v}(x, v, t) = 0.$$

Bolzano-Weierstrass theorem (for the real line) If $A \subset \mathbb{R}$ is infinite and bounded, then there exists at least one point $x \in \mathbb{R}$ that is an *accumulation* point of A; equivalently every bounded sequence in $\mathbb{R}$ has a *convergent subsequence*.

In metric spaces: compactness and sequential compactness are equivalent.

bond There is a chemical bond between two atoms or groups of atoms in the case that the forces acting between them are such as to lead to the formation of an aggregate with sufficient stability to make it convenient for the chemist to consider it as an independent "molecular species."

See also *coordination*.

bond order, p_{rs} The theoretical index of the degree of bonding between two atoms relative to that of a single bond, i.e., the bond provided by one localized electron pair. In molecular orbital theory it is the sum of the products of the corresponding atomic orbital coefficients (weights) over all the occupied molecular spin-orbitals.

Borel sets The sigma-algebra of *Borel sets* of $\mathbb{R}^n$ is generated by the open sets of $\mathbb{R}^n$. An element of this algebra is called *Borel measurable*.

Bose-Einstein gas A gas composed of particles with integral spin.

Bose-Einstein statistics In quantum statistics of the distribution of particles among various possible energy values there are two types of particles, *fermions* and *bosons*, which obey the *Fermi-Dirac statistics* and *Bose-Einstein statistics*, respectively. In the *Fermi-Dirac statistics*, no more than one set of identical particles may occupy a particular quantum state (i.e., the Pauli exclusion principle applies), whereas in the *Bose-Einstein statistics* the occupation number is not limited in any way.

Boson A particle described by *Bose-Einstein statistics*.

boundary Let $A \subset S$ be *topological spaces*. The *boundary* of A is the set $\partial A = \bar{A} - A^\circ$, where $\bar{A}$ is the *closure* and A° is the *interior* of A in S.

boundary layer The motion of a fluid of low viscosity (e.g., air, water) around (or through) a stationary body possesses the free velocity of an ideal fluid everywhere except in an extremely thin layer immediately next to the body, called the *boundary layer*.

boundary value problem The problem of finding a solution to a given differential equation in a given set A with the solution required to meet certain specified requirements on the *boundary* ∂A of that set.

bounded linear operator A *bounded linear operator* from a *normed* linear space $(X_1, \|.\|_1)$ to another *normed* linear space $(X_2, \|.\|_2)$ is a map $T : X_1 \to X_2$ which satisfies

(i.) $T(\alpha x + \beta y) = \alpha T(x) + \beta T(y)$ for all $x, y \in X_1, \alpha, \beta \in \mathbb{R}$; linearity

(ii.) $\|Tx\|_2 \leq C\|x\|_1$, for some constant $C \geq 0$, all $x \in X_1$; boundedness.

Boundedness is equivalent to continuity.

Bourbaki, N. A pseudonym of a changing group of leading French mathematicians. The Association of Collaborators of Nicolas Bourbaki was created in 1935. With the series of monographs *Eléments de Mathématique* they tried to write a foundation of mathematics based on simple structures.

Boussinesq equation

$$u_{tt} - u_{xx} + 3(u^2)_{xx} - u_{xxxx} = 0.$$

Bramble-Hilbert lemma   A crucial tool for proving local interpolation estimates for *parametric equivalent finite elements*.   Given a bounded domain $\Omega \subset \mathbb{R}^n$ with Lipschitz-continuous boundary, it states that, for $k \in \mathbb{N}_0$, $0 < s \leq k + 1$,

$$\inf_{p \in \mathcal{P}_k(\Omega)} \|u - p\|_{H^s(\Omega)} \leq C|u|_{H^s(\Omega)}$$

$$\forall u \in H^s(\Omega),$$

where $C = C(s, k, \Omega) > 0$ and $\mathcal{P}_k(\Omega)$ stands for the space of multivariate polynomials of total degree $\leq k$ on Ω. The proof of this lemma relies on the compact embedding of $H^s(\Omega)$ in $L^2(\Omega)$, a fact that is known as Rellich's lemma. The *Bramble-Hilbert lemma* can be extended to spaces of *polynomials* with separate degree $\leq k$ in each independent variable and to non-standard anisotropic Sobolev spaces.

branch point (in polymers)   A point on a *chain* at which a *branch* is attached.

Notes: **(1)** A branch point from which f linear chains emanate may be termed an *f-functional branch point*, e.g., five-functional branch point. Alternatively, the terms trifunctional, tetrafunctional, pentafunctional, etc. may be used, e.g., pentafunctional branch point.

(2) A branch point in a network may be termed a *junction point*.

branched chain (in polymers)   A *chain* with at least one *branch point* intermediate between the boundary units.

branching process   A stochastic process, X_n, models the number of individuals in the nth generation. Usually both X and n take integer values, and X_n is Markovian. Let Z be the random variable representing the number of offspring in the next generation of a single individual. Assuming all individuals are identical, then X_{n+1} is the sum of the X_n values of Z. This sum of a random number of identically, independent random variables can be analytically obtained using the method of the generating function: $Q_{n+1}(s) = Q_n(R(s))$ where generating functions $Q_n(s) = \sum_{k=0}^{\infty} Prob\{X_n = i\}s^k$ and $R(s) = \sum_{k=0}^{\infty} Prob\{Z = k\}s^k$.

bridge   An edge spanning two connected components of a *graph*.

Comment: The mathematical word accurately conveys the structure: two subgraphs joined by a single edge.

Brouwer's fixed point theorem   Every *continuous* map from the closed ball in $\mathbb{R}^n$ into itself has at least one *fixed point*.

Brownian dynamics   To circumvent the difficulties of MD (see *molecular dynamics*), one approach is to introduce a larger time step in the simulation. This naturally leads to motion being stochastic and hence the dynamic of the molecule is Brownian motion–like. One essential difference between *Brownian dynamics* and MD is that the water molecules are explicit in the latter, whereas they contribute as a random force and a viscous medium in the former. *Brownian dynamics* have the advantage of large time steps. Its diffculty lies in the uncertainty about the interatomic interaction on the large time step and in implicit water, known as coarse-graining.

Brownian motion   A stochastic process $\{X(t) : t \geq 0\}$ is a *Brownian motion process* (or *Wiener process*) if

(i.) $X(0) = 0$;

(ii.)   for each t, $X(t)$ is a *normal* random variable with zero mean;

(iii.)   if $a < b \leq c < d$, the random variables $X(b) - X(a)$ and $X(d) - X(c)$ are independent and have the same distributions whenever $b - a = d - c$.

BRST   Named after *Becchi, Rouet, Stora, and Tyutin*, BRST quantization is a method of quantizing gauge theories. After the introduction of ghost fields the effective *Lagrangian* is no longer gauge invariant, but has a new global symmetry, called *BRST* symmetry. The *BRST* operator **s** is defined on the algebra of local operators making it into a differential graded algebra. The induced *coboundary operator* of the associated *cohomology* (called *BRST cohomology*) is the *BRST operator* **s**, satisfying $\mathbf{s}^2 = 0$. The classical BRST transformations of the *vector* potential A and the ghost field η are

$$\mathbf{s}A = d\eta + [A, \eta], \quad \mathbf{s}\eta = -\frac{1}{2}[\eta, \eta].$$

bulk sample   The sample resulting from the planned aggregation or combination or *sample units*.

bundle   A triple (E, B, π), consisting of two *topological spaces* E and B and a *continous surjective* map $\pi : E \to B$. E is called the total space, B the base space, and $\pi^{-1}(x)$ the fiber at $x \in B$. A *trivial bundle* is of the form $\pi : B \times E \to B$ with $\pi = pr_1$ the projection onto the first factor. If all fibers $\pi^{-1}(x)$ are *homeomorphic* to a space F, and the *bundle* is locally trivial; i.e., there are *homeomorphisms* $\psi_i : U_i \times F \to \pi^{-1}(U_i)$ (U_i an open cover of B) with transition maps which are *homeomorphisms*, then the *bundle* is called a *fiber bundle* and F is called the *typical fiber*. If the typical fiber F is a *vector* space, the *bundle* is called a *vector bundle*. If the spaces E, B are smooth *manifolds* and all the maps above are *smooth*, then the *bundles* are called *smooth fiber bundles*. Examples are the *tangent* bundle and *cotangent* bundle of a *smooth manifold* B. A *principal fiber bundle* consists of a *smooth fiber bundle* (E, B, π) and a *Lie group* G acting freely on E on the right $(u, g) \in E \times G \to ug \in E$ satisfying $\psi_i(x, gh) = \psi_i(x, g)h$, $x \in U_i, g, h \in G$.

bundle morphisms   If $\mathcal{B} = (B, M, \pi, F)$ and $\mathcal{B}' = (B', M', \pi', F')$ are two *fiber bundles*, a *bundle morphism* is a pair of maps (Φ, ϕ) such that $\Phi : B \to B'$, $\phi : M \to M'$ and $\pi' \circ \Phi = \phi \circ \pi$; i.e., Φ sends fibers into fibers. One usually summarizes this property by saying that the following diagram:

$$
\begin{array}{ccc}
B & \xrightarrow{\Phi} & B' \\
\downarrow & & \downarrow \\
M & \xrightarrow{\phi} & M'
\end{array}
$$

is *commutative*. If the *bundles* are endowed with some additional structure (e.g., *vector bundles*) the *bundle morphism* is usually required to preserve that structure (e.g., to be linear on fibers).

A *bundle morphism* (Φ, ϕ) is called a *strong morphism* if ϕ is a *diffeomorphism*. It is called *vertical* if $M = M'$ and $\phi = \mathrm{id}_M$. If Φ is *surjective*, (Φ, ϕ) is called a *bundle epimorphism*. If Φ is *injective*, (Φ, ϕ) is called a *bundle monomorphism*. If Φ is a *diffeomorphism*, then ϕ is also a *diffeomorphism* and (Φ^{-1}, ϕ^{-1}) is a *bundle morphism*, called the *inverse morphism*. In that case (Φ, ϕ) is called a *bundle isomorphism*.

If (x^μ, y^i), (x'^μ, y'^i) are *fibered coordinates* on $\mathcal{B}$ and $\mathcal{B}'$, the local expression of a fibered morphism is:

$$
\begin{cases}
x'^\mu = f^\mu(x) \\
y'^i = Y^i(x, y)
\end{cases}
$$

Burger's equation $u_t = u_{xx} + u_x^2$.

bursting   Some biological cells exhibit brief bursts of oscillations in their membrane electric potential interspersed with quiescent periods during which the membrane potential changes only slowly. The first mathematical model for this phenomenon was proposed by T.R. Chay and J. Keizer in terms of five coupled nonlinear ordinary differential equations in which a slow oscillator modulates a high frequency oscillation. When the slow oscillating variable (S) passes some numerical value μ, the high frequency oscillation occurs; while when the slow oscillating variable is below the critical value, the fast oscillation disappears and the corresponding variable F changes slowly. Hence the overall dynamics of F shows bursts of high frequency oscillations when $S > \mu$, interspered with quiescent periods when $S < \mu$. (*cf.* J. Rinzel, Burst oscillations in an excitable membrane model. In: *Ordinary and Partial Differential Equations*, Eds. B.D. Sleeman and R.J. Jarvis, Springer-Verlag, New York, 1985).

C

Calabi-Yau spaces   Complex spaces with a vanishing first *Chern class* or, equivalently, with trivial canonical bundle (canonical class). They are used to construct possibly realistic (super) string models.

calcium-induced calcium release   A positive feedback component in the calcium dynamics of biological cells (see *autocatalysis*). It is known from experiments that some calcium channels (see *ion channel*) which are responsible for calcium influx into cytosol are positively modulated by the calcium in the cytosol. This mechanism has been suggested as being responsible for the widely observed calcium oscillation in cells.

calibration curve   See *calibration function*.

calibration function (in analysis)   The functional (not statistical) relationship for the *chemical measurement process*, relating the expected value of the observed (gross) signal or response variable $E(y)$ to the analyte amount x. The corresponding graphical display for a single analyte is referred to as the *calibration curve*. When extended to additional variables or analytes which occur in multicomponent analysis, the "curve" becomes a calibration surface or hypersurface.

Callan-Symanzik equation   A type of renormatization group equation in quantum field theory which studies the variation of the Green's function with respect to the physical mass.

Camassa-Holm equation   A shallow water equation which is a *completely integrable system* having peakon solutions

$$\partial_t u - \partial_t \partial_x^2 u + 3u\partial_x u - 2\partial_x u \partial_x^2 u - u\partial_x^3 u = 0.$$

Campbell-Hausdorff formula   Also known as *Baker-Campbell-Hausdorff formula*, is the formula for the product of exponentials in a Lie algebra $\exp A \exp B = \exp\{A + B + \frac{1}{2}[A, B] + \frac{1}{12}[A, [A, B]] + \frac{1}{12}[B, [B, A]] + \cdots + c_n[A[A \ldots [A, B] \ldots]]\}$.

candidate device   A subgraph of the biochemical graph with particular topological properties, without necessarily having distinct biochemical or dynamical properties.

canonical bracket   On a *symplectic manifold* (M, ω) with *canonical coordinates* $(q^1, \ldots q^n, p_1, \ldots p_n)$ we have the *canonical (Poisson) bracket* between any two functions F and G on M as

$$\{F, G\} = \sum_{i=1}^{n} \left(\frac{\partial F}{\partial q^i} \frac{\partial G}{\partial q_i} - \frac{\partial F}{\partial p_i} \frac{\partial G}{\partial q^i} \right).$$

canonical coordinates   (Darboux's theorem) On any *symplectic manifold* (M, ω) there exist local coordinates $(q^1, \ldots q^n, p_1, \ldots p_n)$, called *canonical coordinates*, such that

$$\omega = \sum_{i=1}^{n} dq^i \wedge dp_i .$$

canonical one-form   On any *cotangent manifold* T^*M there is a unique one-form Θ such that $\alpha^*\Theta = \alpha$ for any one form α on M. In *canonical coordinates* $(q^1, \ldots q^n, p_1, \ldots p_n)$, Θ is given by

$$\Theta = \sum_{i=1}^{n} p_i dq^i.$$

canonical symplectic form   On any *cotangent manifold* T^*M there is a unique symplectic form Ω defined as $\Omega = -d\Theta$ where Θ is the *canonical one-form*. In *canonical coordinates* $(q^1, \ldots q^n, p_1, \ldots p_n)$, Ω is given by

$$\Omega = \sum_{i=1}^{n} dq^i \wedge dp_i.$$

canonical transformation   A *smooth* map f between two *symplectic manifolds* (M_1, ω_1) and (M_2, ω_2) is called *canonical* or *symplectic* if it preserves the symplectic forms, i.e., $f : (M_1, \omega_1) \to (M_2, \omega_2)$, $f^*\omega_2 = \omega_1$.

cardinality The number of elements in a set.

Comment: Often denoted by $\overline{S}$ or $|S|$, where S is any set. See also *countable set, denumerably infinite set, finite set, infinite set*, and *uncountably infinite set*.

carotenes *Hydrocarbon carotenoids* (a subclass of tetraterpenes).

carotenoids Tetraterpenoids (C_{40}), formally derived from the acyclic parent ψ, ψ-carotene I by hydrogenation, dehydrogenation, cyclization, oxidation, or combination of these processes. This class includes *carotenes, xanthophylls*, and certain compounds that arise from rearrangement of the skeleton of I or by loss of part of this structure. *Retinoids* are excluded.

Cartan matrix For a semisimple *Lie algebra* of rank l the *Cartan matrix* $[A_{ij}]$ is defined by $A_{ij} = \frac{2(\alpha_i, \alpha_j)}{(\alpha_i, \alpha_i)}$ where $\alpha_1, \cdots, \alpha_l$ is the system of simple roots.

Casimir Elements in the universal enveloping algebra of a *Lie algebra* that commute with all other elements of the *Lie algebra* are called *Casimirs*.

Casimir function On a *Poisson manifold* $(P, \{\ ,\ \})$, a function C that Poisson commutes with every function F is called a *Casimir function*, i.e., $\{C, F\} = 0$ for all F.

Casorati-Weierstrass theorem Let f have an *essential singularity* at z_0 and let $w \in$ C. Then there exists $z_1, z_2, z_3, \ldots$ such that $z_n \to z_0$ and $f(z_n) \to w$.

catalyst A substance that increases the rate of a reaction without modifying the overall standard Gibbs energy change in the reaction; the process is called catalysis. The *catalyst* is both a *reactant* and *product* of the reaction. The words *catalyst* and catalysis should not be used when the added substance reduces the rate of reaction (see *inhibitor*).

Catalysis can be classified as homogeneous catalysis, in which only one phase is involved, and heterogeneous catalysis, in which the reaction occurs at or near an interface between phases. Catalysis brought about by one of the products of a reaction is called autocatalysis. Catalysis brought about by a group on a reactant molecule itself is called intramolecular catalysis.

The term catalysis is also often used when the substance is consumed in the reaction (for example: base-catalyzed hydrolysis of esters). Strictly, such a substance should be called an activator.

Comment: Reactions are written at many levels of resolution, and those reactions which detail how the *catalyst* functions will not meet this definition. But that is appropriate, since the molecule which functions catalytically in one reaction will be a substrate in the reactions which describe the catalysis.

Cauchy-Goursat theorem Suppose that $f : D \to$ C is *analytic* on a disk $D \subset$ C; then f has an antiderivative F on D, i.e., $F'(z) = f(z)$, which is analytic in D, and, if γ is any *closed curve* in D, then $\int_\gamma f = 0$.

Cauchy integral formula Let f be *analytic* within and on a simple closed curve γ and z_0 a point in the interior of γ. Then

$$f(z_0) = \frac{1}{2\pi i} \int_\gamma \frac{f(z)}{z - z_0} dz.$$

Cauchy-Riemann equations (Cauchy-Riemann theorem) Let $\Omega \subset$ C be an open set and $f : \Omega \to$ C be a function $f(z) = u(x, y) + iv(x, y)$. Then $f'(z_0)$ exists if and only if f is differentiable in the sense of real variables and, at $z_0 = (x_0, y_0)$, u, v satisfy the *Cauchy-Riemann equations*

$$\frac{\partial u}{\partial x} = \frac{\partial v}{\partial y}, \quad \text{and} \quad \frac{\partial u}{\partial y} = -\frac{\partial v}{\partial x}.$$

Cauchy-Riemann theorem See *Cauchy-Riemann equations*.

Cauchy-Schwarz inequality In any *inner product* space $(V, <\ ,\ >)$ for any $x, y \in V$

$$|<x, y>| \leq \|x\| \|y\|.$$

Cauchy sequence A sequence $\{x_n\}$ in a *metric space* (M, d) such that for every $\epsilon > 0$ there exists an integer N such that $d(x_m, x_n) < \epsilon$ whenever $m > N$ and $n > N$.

Cauchy theorem   If a function f is *analytic* in a region Ω and γ is a closed curve in Ω which is homotopic to a point in Ω, then $\int_\gamma f = 0$. See *homotopy*.

Cayley transform   Let A be a *closed symmetric operator* on a *Hilbert space*. Then $(A - iI)(A + iI)^{-1}$ is called the *Cayley transform* of A.

Cea's lemma   Let a be a bilinear/sesquilinear form on a real/complex *Banach space* V, which satisfies:

Continuity:
$$|a(u, v)| \leq C\|u\|_V \|v\|_V, \quad \forall u, v \in V,$$

V-*ellipticity*:
$$|\mathcal{R}\{a(u, v)\}| \geq \alpha \|u\|_V^2, \quad \forall u \in V.$$

Suppose that $V_h \subset V$ is a closed subspace of V. For $f \in V'$, let $u \in V, u_h \in V_h$ stand for solutions of the variational problems

$$a(u, v) = f(v) \quad \forall v \in V,$$

$$a(u_h, v_h) = f(v_h) \quad \forall v_h \in V_h.$$

These are unique according to the *Bramble-Hilbert lemma*. Then *Cea's lemma* asserts that

$$\|u - u_h\| \leq \frac{C}{\alpha} \inf_{v_h \in V_h} \|u - v_h\|_V.$$

If a is symmetric/Hermitian the constant $\frac{C}{\alpha}$ can be replaced with 1. A generalization of *Cea's lemma* to sesqui-linear forms that satisfy an *inf-sup condition* is possible. Assume that instead of being V-elliptic the bilinear form a satisfies, with $\alpha > 0$,

$$\sup_{v_h \in V_h} \frac{|a(u_h, v_h)|}{\|v_h\|_V} \geq \alpha \|u_h\|_V \quad \forall u_h \in V_h.$$

Then, provided that a unique continuous solution $u \in V$ of the above variational problem exists, there holds

$$\|u - u_h\|_V \leq \left(1 + C\alpha^{-1}\right) \inf_{v_h \in V_h} \|u - v_h\|_V.$$

chain (in polymers)   The whole or part of a *macromolecule*, an *oligomer molecule* or a *block*, comprising a linear or branched sequence of *constitutional units* between two boundary constitutional units, each of which may be either an *end-group*, a *branch point* or an otherwise-designated characteristic feature of the *macromolecule*.

Notes: (**1**) Except in linear single-strand *macromolecules*, the definition of a *chain* may be somewhat arbitrary.

(**2**) A cyclic *macromolecule* has no end groups but may nevertheless be regarded as a *chain*.

(**3**) Any number of branch points may be present between the boundary units.

(**4**) Where appropriate, definitions relating to *macromolecule* may also be applied to *chain*.

chain rule   Let X, Y, Z be *Banach spaces* and $f : X \to Y$, $g : Y \to Z$ be *differentiable* of class C^k. Then $g \circ f : X \to Z$ is of class C^k and

$$D(g \circ f)(x) = Dg(f(x)) \circ Df(x).$$

See *derivative*.

characteristic classes   *Chern classes* $c_1, \ldots, c_k$ are defined for a complex *vector bundle* of dimension k (or equivalently for a $GL(k, \mathbf{C})$ *principal bundle*) $c_i \in H^{2i}(M)$.

Pontrjagin classes $p_1, \ldots, p_j$ are defined for a real *vector bundle* of dimension k (or equivalently for a $GL(k, \mathbb{R})$ *principal bundle*) $p_i \in H^{4i}(M)$.

Stiefel-Whitney classes $w_1, \ldots, w_k$ are defined for a *real vector bundle* of dimension k (or equivalently for a $GL(k, \mathbb{R})$ *principal bundle*). They are $\mathbf{Z}_2$ characteristic classes $w_i \in H^i(M; \mathbf{Z}_2)$.

characteristic cone   The principal symbol $P_m(x, \xi) = \sum_{|\alpha|=m} c_\alpha(x)\xi^\alpha$ of a (linear partial) differential operator $P(x, D) = \sum_{|\alpha|\leq m} c_\alpha(x)D^\alpha$ is homogeneous of degree m in ξ. The set

$$C_{P_m}(x) = \{\xi \in \mathbb{R}^n \mid P_m(x, \xi) = 0\}$$

is called the *characteristic cone* of P at x.

characteristic equation   For an $n \times n$ *matrix* A the equation $\det(A - \lambda I) = 0$.

characteristic function   The *characteristic function* of a set A is defined by

$$\chi_A(x) = \begin{cases} 1 & \text{if } x \in A \\ 0 & \text{if } x \notin A \end{cases}$$

characteristic polynomial   See *characteristic equation*.

characteristic set   The *characteristic set* of a *pseudodifferential operator* P is the subset $Char(P) \subset T^*M - 0$ defined by $p(x, \xi) = 0$, where $p(x, \xi) : T^*M - 0 \to \mathbb{R}$ is the principal symbol of P.

characteristic x-ray emission   X-ray emission originates from the radiative decay of electronically highly excited states of matter. Excitation by electrons is called primary excitation and by photons, secondary or fluorescence excitation.   Particle induced x-ray emission (PIXE) is produced by the excitation of heavier particles such as protons, deuterons, or heavy atoms in varying degrees of ionization. Emission of photons in the x-ray wavelength region occurs from ionized gases or plasmas at high temperatures, nuclear processes (low-energy end of the gamma-ray spectrum) and from radiative transitions between muonic states.

Characteristic x-ray emission consists of a series of x-ray spectral lines with discrete frequencies, characteristic of the emitting atom. Other features are emission bands from transitions to valence levels. In a spectrum obtained with electron or photon excitation the most intense lines are called diagram lines or normal x-ray lines. They are dipole allowed transitions between normal x-ray *diagram levels*.

charge conjugation   For a *Dirac spinor* $\psi = \begin{pmatrix} \psi_L \\ \psi_R \end{pmatrix}$ the charge conjugate spinor is $\psi^c = \begin{pmatrix} \sigma^2 \psi_R^* \\ -\sigma^2 \psi_L^* \end{pmatrix}$, where $*$ means the *complex conjugate* and $\sigma^2 = \begin{pmatrix} 0 & -i \\ i & 0 \end{pmatrix}$ is the secomd Pauli spin matrix. Note that $(\psi^*)^* = \psi$.

chart   A *chart* on a *manifold* M is a pair (U, ϕ) where U is an open subset of M and ϕ is *bijection* form U onto an open subset of $\mathbb{R}^n$. U is called a *coordinate patch*, $\phi_i \phi_j^{-1}$ are *coordinate transition functions*.

Chebyshev's inequality   If $f \in (L^p, \mu)$ $(0 < p < \infty)$, then for any $\alpha > 0$

$$\mu\{x : |f(x)| > \alpha\} \leq \left(\frac{\|f\|_p}{\alpha} \right)^p.$$

chemical equation   A balanced elementary reaction equation, bimolecular or less on one side, which describes the organic chemistry of the noncatalytic species of an overall biochemical reaction by one of five fundamental organic mechanisms (substitution, addition, elimination, rearrangement, and oxidation-reduction). Equally, a set of such reaction equations which sums to the overall biochemical reaction.

Comment: These mechanisms distinguish all the various subtypes (nucleophilic, electrophilic, free-radical; heterolytic, homolytic, pericyclic; direct electron, hydride, hydrogen atom transfer, ester intermediates, displacement, and addition-elimination reactions).

chemical measurement process (CMP)   An analytical method of defined structure that has been brought into a state of statistical control, such that its imprecision and bias are fixed, given the measurement conditions.   This is prerequisite for the evaluation of the performing characteristics of the method, or the development of meaningful uncertainty statements concerning analytical results.

chemical motif   A *motif* describing a chemical relationship between two compounds in the donor-acceptor formalism.

Comment: The constraint for bimolecular relationship permits use of the common donor-acceptor language. A reaction may have more than one such relationship. See also *biochemical, dynamical, functional, kinetic, mechanistic, phylogenetic, regulatory, thermodynamic,* and *topological motives.*

chemical potential   In thermodynamics, the partial molar Gibbs free energy. In a thermal equilibrium with many chemical reactions, all components have to have the same chemical potential.

chemical reaction   A process that results in the interconversion of *chemical species*. Chemical reactions may be *elementary reactions* or *stepwise reactions*. (It should be noted that this definition includes experimentally observable interconversions of conformers.)

Detectable chemical reactions normally involve sets of *molecular entities*, as indicated by this definition, but it is often conceptually convenient to use the term also for changes involving single molecular entities (i.e., "microscopic chemical events").

chemical species   An ensemble of chemically identical *molecular entities* that can explore the same set of molecular energy levels on the time scale of the experiment. The term is applied equally to a set of chemically identical atomic or molecular structural units in a solid array.

For example, two conformational *isomers* may be interconverted sufficiently slowly to be detectable by separate NMR spectra and hence to be considered to be separate chemical species on a time scale governed by the radiofrequency of the spectrometer used. On the other hand, in a slow chemical reaction the same mixture of conformers may behave as a single chemical species, i.e., there is virtually complete equilibrium population of the total set of molecular energy levels belonging to the two conformers. Except where the context requires otherwise, the term is taken to refer to a set of molecular entities containing isotopes in their natural abundance. The wording of the definition given in the first paragraph is intended to embrace both cases such as graphite, sodium chloride, or a surface oxide, where the basic structural units may not be capable of isolated existence, as well as those cases where they are.

In common chemical usage generic and specific chemical names (such as *radical* or hydroxide ion) or chemical formulae refer either to a chemical species or to a *molecular entity*.

Chern classes   See *characteristic classes*.

chiral group   In QCD (quantum chromo dynamics) the group $SU_3 \times SU_3$. See *chromodynamics, quantum*.

chiral transformation   The transformation of a *Dirac spinor* $\psi \to e^{i\beta\gamma_5}\psi$ is called *chiral transformation*, or *chiral symmetry*, where β is a constant and $\gamma_5 = i\gamma^0\gamma^1\gamma^2\gamma^3$ with the 4×4 gamma matrices defined by the Pauli matrices

$$\sigma^1 = \begin{pmatrix} 0 & 1 \\ 1 & 0 \end{pmatrix}, \sigma^2 = \begin{pmatrix} 0 & -i \\ i & 0 \end{pmatrix}, \sigma^3 = \begin{pmatrix} 1 & 0 \\ 0 & -1 \end{pmatrix}$$

by $\gamma^i = \begin{pmatrix} 0 & -\sigma^i \\ \sigma^i & 0 \end{pmatrix}$.

chirality   The components ψ_L, ψ_R of a *Dirac spinor* $\psi \to e^{i\beta\gamma_5}\psi$ are called *chiral components* or *Weyl components*.

cholesteric phase   See *liquid-crystal transitions*.

Christoffel symbols   on a *(pseudo)-Riemannian manifold* (M, g), the coefficients of the *Levi-Civita connection*. If $g = g_{\mu\nu}(x)\,dx^\mu \otimes dx^\nu$ is the local expression of the metric tensor, its *Christoffel symbols* are given by:

$$\{^\alpha{}_{\beta\mu}\}_g = \tfrac{1}{2}g^{\alpha\epsilon}\left(-\partial_\epsilon g_{\beta\mu} + \partial_\beta g_{\mu\epsilon} + \partial_\mu g_{\epsilon\beta}\right)$$

where $g^{\alpha\epsilon}$ denotes the *covariant inverse metric*, and ∂_μ denotes the partial derivative with respect to x^μ.

Under changes of local coordinates $x'^\mu = x'^\mu(x)$, *Christoffel symbols* transform as:

$$\{^\alpha{}_{\beta\mu}\}'_g = J^\alpha_\gamma\left(\{^\gamma{}_{\delta\nu}\}_g \bar{J}^\delta_\beta \bar{J}^\nu_\mu + \bar{J}^\gamma_{\beta\mu}\right)$$

where $J^\alpha_\gamma = \frac{\partial x'^\alpha}{\partial x^\gamma}$ is the jacobian of the coordinate transformation, $\bar{J}^\gamma_\alpha = \frac{\partial x^\gamma}{\partial x'^\alpha}$ is the inverse Jacobian, and we set $\bar{J}^\gamma_{\beta\mu} = \frac{\partial^2 x^\gamma}{\partial x'^\beta \partial x'^\mu}$. If ∇ denotes the *covariant derivative operator* associated to the *Levi-Civita connection*, then one has:

$$\nabla_{\partial_\mu}\partial_\nu = \{^\lambda{}_{\mu\nu}\}_g\,\partial_\lambda.$$

chromatic number   The minimum number of colors needed to color the nodes of a graph, such that no two adjacent nodes have the same color. Denoted $\chi(\mathsf{G})$.

chromodynamics, quantum (QCD)   The quantum field theory describing the strong interactions of quarks and gluons.

chromophore   The part (atom or group of atoms) of a *molecular entity* in which the electronic transition responsible for a given spectral band is approximately localized. The term arose in the dyestuff industry, referring originally to the groupings in the moleculaer that are responsible for the dye's color.

chromosome A self-replicating structure consisting of *DNA* complexes with various proteins and involved in the storage and transmission of genetic information; the physical structure that contains *genes* (*cf. plasmid*). Eukaryotic cells have a characteristic number of *chromosomes* per cell (*cf. ploidy*) and contain DNA as linear duplexes. The chromosomes of bacteria consist of double-standed, circular DNA molecules.

circulation In the theory of Markov processes, the probability flux in a stationary state which can be achieved by two types of balance: detailed balance and circular balance. For most Markov processes, the sufficient and necessary condition for zero circulation is time-reversibility.

class C^k A function $f : U \subset \mathbb{R}^n \to \mathbb{R}^m$ is differentiable of *class C^k*, $0 \leq k \leq \infty$, if all partial derivatives $\frac{\partial^\alpha f}{\partial x^\alpha}$, $0 \leq |\alpha| \leq k$, of f up to order k exist and are continuous.

classical Fourier integral operator A Fourier integral operator $Au = \int \int e^{\phi(x,\xi)} a(x, \xi)u(y)dy$ is called *classical* if it has a *classical symbol* $a(x, \xi)$.

classical groups The *matrix* groups $GL(n)$, $SL(n)$, $U(n)$, $SU(n)$, $O(n)$, $SO(n)$, $Sp(2n)$ are called *classical Lie groups*.

classical limit In quantum mechanics when the Planck constant $h \to 0$ classical mechanics is recovered.

classical mechanics Classical mechanics vis-à-vis quantum mechanics and relativistic mechanics. The equations of motion in classical mechanics are *Hamilton's equations* or equivalently the *Euler-Lagrange equations*, formulated on a finite dimensional *symplectic manifold*.

classical path A distinction between classical and quantum mechanics. In classical mechanics a particle takes only one path to go from a point q to a point q', while all paths contribute in quantum mechanics. The three colors R, G, and B belong to the *representation* of $SU(3)$.

classical symbol A symbol $a(x, \xi)$ of a *pseudodifferential operator* (or *Fourier integral operator*) of order m is called *classical* if there exist C^∞ functions $a_j(x, \xi)$, positively homogeneous of degree $m - j$ in ξ (i.e., $a(x, \tau\xi) = \tau^{m-j}a_j(x, \xi)$, $\tau > 0$) such that asymptotically

$$a(x, \xi) \sim \sum_{j=0}^{\infty} a_j(x, \xi).$$

Clifford algebra *Clifford algebra* is a formulation of algebra which unifies and extends complex numbers and vector algebra. It is based on the Clifford product of two vectors a and b which is written ab. The product has two parts, a scalar part and a bivector part. The scalar part is symmetric and corresponds with the usual dot product $a \cdot b = \frac{1}{2}(ab + ba)$. The bivector part is antisymmetric and can be thought of as a directed area, defining a plane $a \wedge b = \frac{1}{2}(ab - ba)$. Then the Clifford product can be written as $ab = a \cdot b + a \wedge b$.

closed curve A *closed curve* (or *closed path*) in a space M is a *curve* $\gamma : [a, b] \to M$ such that $\gamma(a) = \gamma(b)$.

closed form An *exterior form* ω on a *manifold* having vanishing *exterior differential* $d\omega = 0$. *Exact forms* are closed. The converse is not true in general. It is true on contractible *manifolds* (e.g., on *star-shaped* open sets of $\mathbb{R}^m$ by Poincaré's lemma). *De Rham cohomology* studies the topological properties of *manifolds* by classifying closed forms that are not exact.

Example: The form $\omega = (x^2 + y^2)^{-1}(xdy - ydx)$ is a *closed* (but not exact) form defined on $\mathbb{R}^2 - \{0\}$; in fact, it locally reduces to $d\varphi$ on the unit circle $x^2 + y^2 = 1$, but it is not an exact form since φ cannot extend to a single-valued coordinate function on the whole of the circle.

closed graph theorem If X and Y are *Banach spaces* and $T : X \to Y$ is a *closed* linear operator defined on all of X then T is *bounded* (i.e., *continuous*).

closed operator A *linear operator* $T : X \to Y$, between two *Banach spaces* X and Y and having a dense domain $D(T) \subset X$ is *closed* if its graph $\Gamma(T) = \{(x, Tx) : x \in D(T)\}$ is closed in $X \times Y$.

closed orbit   An orbit $\gamma(t)$ of a *vector field* X is called *closed* if there is a $\tau > 0$ such that $\gamma(t + \tau) = \gamma(t)$ for all t.

closed set   The complement of an *open set* in a *topological space* $(X, \tau(X))$. The class $\chi(X)$ of closed sets has the following properties:

(i.)   The empty set $\emptyset$ and the whole space X are elements in $\chi(X)$;

(ii.)   The union of a finite number of elements in $\chi(X)$ is still in $\chi(X)$; and

(iii.)   The intersection of a (possibly infinite) family of elements in $\chi(X)$ is still in $\chi(X)$.

closure   For a subset A of a *topological space*, the smallest closed set containing A. Denoted $\bar{A}$.

coacervation   The separation into two liquid phases in *colloidal* systems. The phase that is more concentrated in *colloid* component is the coacervate, and the other phase is the equilibrium solution.

coadjoint action   The *coadjoint action* Ad^* of a *Lie group* G on the dual of its *Lie algebra* $\mathbf{g}^*$ is defined as the dual of the adjoint action of G on $\mathbf{g}$. It is given by

$$< Ad_g^*\alpha, \xi > = < \alpha, Ad_g\xi >,$$

for $g \in G$, $\alpha \in \mathbf{g}^*$, $\xi \in \mathbf{g}$, where $<,>$ is the pairing between $\mathbf{g}^*$ and $\mathbf{g}$ and $Ad_g(\xi) = T_e(R_{g^{-1}} \circ L_g)\xi$.

coadjoint orbit   The orbits of the *coadjoint action*, i.e., $\mathcal{O}_\alpha = \{Ad_g^*\alpha \mid g \in G, \alpha \in \mathbf{g}^*\} \subset \mathbf{g}^*$. *Coadjoint orbits* carry a natural *symplectic* structure, the induced *Poisson bracket* is the *Lie-Poisson bracket* given by

$$\{F, H\}(\alpha) = -\left\langle \alpha, \left[\frac{\delta F}{\delta \alpha}, \frac{\delta H}{\delta \alpha}\right]\right\rangle.$$

coadjoint representation   The *coadjoint representation* of a *Lie group* G on the dual of its *Lie algebra* $\mathbf{g}^*$ is given by the *coadjoint action* Ad^* by

$$Ad^* : G \to GL(\mathbf{g}^*, \mathbf{g}^*),$$
$$Ad_{g^{-1}}^* = (T_e(R_g \circ L_{g^{-1}}))^*.$$

coboundary operator   See *cochain complex*.

cochain complex   A *cochain complex* consists of a sequence of *modules* and *homomorphisms*

$$\cdots \to C^{q-1} \to C^q \to C^{q+1} \to \cdots$$

such that at each stage the image of a given *homomorphism* is contained in the kernel of the next. The *homomorphism* $d^q : C^q \to C^{q+1}$ is called *coboundary operator*. We have $d^2 = 0$. The kernel $Z^q = ker\ d^q$ is the module of qth degree *cocycles* and the image $B^q = im\ d^{q-1}$ is the module of qth degree *coboundaries*. The qth *cohomology module* H^q is defined as $H^q = Z^q/B^q$.

cocycle   See *cochain complex*

codifferential   On a *Riemannian manifold* (M, g) the *codifferential* $\delta : \Omega^k(M) \to \Omega^{k-1}(M)$ is defined by

$$\delta\alpha = (-1)^{n(n+k)+1+Ind(g)} * d * \alpha, \alpha \in \Omega^k(M),$$

where $*$ is the Hodge operator and d the *exterior derivative*. We have $\delta^2 = 0$.

codomain   See *relation*. See also *domain*, *image*, *range*.

coercivity   Let V be a Banach space. A sesqui-linear form $a : V \times V \to \mathbb{C}$ is *coercive* on V, if there exists a constant $\alpha > 0$ and a compact operator $K : V \to V'$ such that

$$|\mathcal{R}\{a(u, u)\} + (K(u), u)| \geq c\|u\|_V^2\ \forall u \in V.$$

cohomology   See *deRham cohomology group*.

coisotropic (sub)manifold   (in a *symplectic manifold* $[P, \omega]$)   A submanifold $M \subset P$ of a *symplectic manifold* (P, ω) such that at any point $p \in M$ the tangent space is contained in its symplectic polar, i.e.,

$$T_pM \subset (T_pM)^o.$$

The dimension of a *coisotropic manifold* M is at least half of the dimension of P.

col (saddle point)   A mountain-pass in a *potential-energy surface* is known as a *col* or *saddle point*. It is a point at which the gradient is zero along all coordinates, and the curvature is positive along all but one coordinate, which is the reaction coordinate, along which the curvature is negative.

colligation   The formation of a covalent bond by the combination or recombination of two *radicals* (the reverse of *unimolecular homolysis*). For example, $\dot{O}H + H_3\dot{C} \rightarrow CH_3OH$.

colloid   A short synonym for *colloidal* system.

colloidal   The term refers to a state of subdivision, implying that the molecules or polymolecular particles dispersed in a medium have at least in one direction a dimension roughly between 1 nm and 1 μm, or that in a system discontinuities are found at distances of that order.

color   (**1**) (in mathematics) A nonlabel token for an edge or node of a graph. A *coloring* of the graph G is the set of colors assigned to the nodes, such that no two adjacent nodes have the same color. An *edge coloring* is the set of colors assigned to edges; a *proper edge coloring* is the set of colors such that no two adjacent edges have the same color.

Comment: These need not be physical colors, though it is perhaps easiest to think of them that way. Note that colorings are by definition proper, whereas edge colorings need not be. The difference between definitions of node and edge colorings may have been motivated in the beginning by the four color map problem.

(**2**) (in quantum QCD) Gluons and quarks have an additional type of polarization (degree of freedom) not related to geometry. "The idiots physicists, unable to come up with any wonderful Greek words anymore, called this type of polarization by the unfortunate name of *color*." (R.P. Feynman in *QED: The Strange Theory of Light and Matter*, Princeton University Press, Princeton, NJ 1985).

commutation relations   If a classical *Hamiltonian* $H(q_i, p_i)$ is quantized, one considers q_i and p_i as operators on a *Hilbert space* satisfying the *commutation relations* $[q_i, p_j] = \delta_{ij}$ and $[q_i, q_j] = [p_i, p_j] = 0$.

commutative   Referring to a set A with a binary operation * satisfying

$$a * b = b * a$$

for all $a, b \in A$. See also *Abelian group*.

commutator   (in an associative algebra A) The binary operation given by

$$[A, B] = A \cdot B - B \cdot A.$$

For example, this is the definition for the commutator of the *Lie algebra* $GL(n, \mathbb{R})$ of $n \times n$ matrices. Notice that it satisfies the *Jacobi identities*

$$[[A, B], C] + [[B, C], A] + [[C, A], B] = 0.$$

The *commutator* of two vector fields $X = X^\mu \partial_\mu$, $Y = Y^\mu \partial_\mu$ is defined by

$$[X, Y] = (X^\mu \partial_\mu Y^\nu - Y^\mu \partial_\mu X^\nu)\partial_\nu.$$

More generally any binary operation defining a *Lie algebra*, namely, the Lie-product obeying Jacobi identities. See also *Lie algebra*.

compact   (**1**) A *topological space* $(X, \tau(X))$ is *compact* if from any open covering of X one can always extract a finite *subcovering*. If X is a topological subspace of a metric space, "compact" is equivalent to "closed and bounded." Thence closed intervals are compact in $\mathbb{R}$; closed balls are compact subsets of $\mathbb{R}^m$ (as well as in any metric space).

(**2**) (locally) A *topological space* X is *locally compact* if every point $p \in X$ has a *compact neighborhood*.

compact operator   Let X and Y be *Banach spaces* and $T : X \rightarrow Y$ a *bounded linear operator*. T is called *compact* or *completely continuous* if T maps bounded sets in X into precompact sets (the closure is compact) in Y. Equivalently, for every bounded sequence $\{x_n\} \subset X$, $\{Tx_n\}$ has a subsequence convergent in Y.

compact set   A subset M of a *topological space* X is called *compact* if every system of open sets of X which covers M contains a finite subsystem also covering M.

compact support   A *continuous function f* has *compact support* if its support $supp(f) = \overline{\{x : f(x) \neq 0\}}$ is a *compact set*.

competition   One class of models for population dynamics in which species are competing for the same, limited resource. Hence, the growth rate of one species decreases with increasing population of the other and vice versa. Such a system is likely to exhibit extinction of the weaker species. Stable co-existence is possible, however, under a delicate balance.

complement   Given a set A, the set of all elements not in A, denoted $\bar{A}$ or A^c.

complete space   A *metric space* (M, d) is *complete* if every *Cauchy sequence* $\{X_n\}$ in M converges in M. That is, there is $x \in M$ such that $d(x_n, x) \to 0$, as $n \to \infty$

complete vector field   A *vector* field X on a *manifold M* is *complete* if its *flow* ϕ_t is defined for all $t \in \mathbb{R}$, $(\frac{d}{dt}\phi_t(x) = X(\phi_t(x))$ for all $t \in \mathbb{R}$).

completely continuous   See *compact operator*.

completely integrable system   A Hamiltonian system defined over a *symplectic manifold* (P, ω) of dimension $2n$ with n *first integrals* F_i in *involution*, i.e., such that $\{F_i, F_j\} = 0$ for all pairs (i, j). The equations of motion of a completely integrable system can be formally integrated. Explicitly, a *Hamiltonian system* (vector field) X_H on $\mathbb{R}^{2n}$ is called *completely integrable* if there exist n constants of the motion $f_1, \ldots f_n$ which are linearly independent and in *involution*, i.e., $\{f_i, f_j\} = 0$ for all $i, j = 1, ..., n$.

complex   A *molecular entity* formed by loose *association* involving two or more component molecular entities (ionic or uncharged), or the corresponding *chemical species*. The bonding between the components is normally weaker than in a covalent *bond*.

The term has also been used with a variety of shades of meaning in different contexts; it is therefore best avoided when a more explicit alternative is applicable. In inorganic chemistry the term "coordination entity" is recommended instead of "complex."

complex conjugate   The *complex conjugate* of the complex number $z = x + iy$ is the complex number $\bar{z} = x - iy$.

complex number   A number of the form $x + iy$, where x and y are real numbers and $i^2 = -1$.

complex structure   A *complex structure* on a real *vector space* V is a linear map $\mathbf{J} : V \to V$ such that $\mathbf{J}^2 = Id$. Setting $iz = \mathbf{J}(z)$ gives V the structure of a *complex vector space*.

complex vector space   A *complex vector space* V is a *vector space* over the field of *complex numbers* C; i.e., scalar multiplication λz is defined for complex number $\lambda \in$ C, $z \in V$.

composition   Consider two functions, $f : A \to B$ and $g : C \to D$, over four sets A, B, C, D. If $f(A) \subseteq C$, then the result of applying f and g in succession is equivalent to the application of a single *composite* function, denoted $g \circ f : A \to D$. $g \circ f$ is defined for any $x \in A$ as $(g \circ f)(x) = g(f(x))$.

compound   Any molecule or assembly of molecules.

Comment: These range in size from single nuclei (H^+) to DNA to transport complexes embedded in cell membranes on up. One often distinguishes between *enzymes* and *metabolites*, smaller molecules. The boundary between "macromolecular" and "smaller" is not fixed precisely, but is perhaps about 1000 daltons.

compound's reactions   A *compound's reactions* is the set of reactions in which that *compound* participates.

Comment: See *reaction's compounds*.

computational biology   The development and application of theory, algorithms, heuristics, and computational systems, including electronic databases, to biological problems; and equally, the application of biological concepts, materials, or processes to problems not originating in biology; and activities at the interface of these two.

Comment: This definition is broad, but still useful; it includes areas such as the representation of information in electronic databases, the construction and maintenance of those databases (including interfaces), the development of biologically realistic automated reasoning methods, and any underlying mathematical techniques, the management of computations over widely distributed, very heterogeneous systems like the World Wide Web; molecular computing; nanotechnology (where computational issues are raised or where biological artifacts are used in computation); and the problems that biology poses for the foundations of computer science.

In a nutshell, computational biology is "computing about biology and computing with biology."

Occasionally, one sees a distinction between computational biology and *bioinformatics*, with the notion that computational biology is theory and algorithms and bioinformatics are databases and retrieval. This is a distinction without a difference, since most biological problems require both. To the extent that biological problems are molecular, computational biology overlaps with computational chemistry. Often the distinction is more one of investigator origin (chemist vs. biologist), degree of resolution of the question (quantum-mechanical vs. tissue), and platform choice (SGI vs. Sun).

computational complexity The time and space an algorithm requires to solve a problem.

Comment: Time and space have a mathematical relationship to the size N and complexity of the problem. Thus algorithms are described as being *linear*, *polynomial*, etc., depending on the function of N in the relation between it and time or space. Functions are designated $O(f(N))$, meaning the algorithm requires on the order of $f(N)$ resources. Algorithms which are *polynomial* represent the practical upper bound of feasibility for large N, with lower values for exponents and determinism of the algorithm obviously preferable. Many problems requiring supra-polynomial resources can be addressed in favorable cases. Occasionally other metrics, such as the number of statements in the program, are defined for particular purposes.

computational step A *computational step* $s_i \in S$ consists of three components:

(i.) recognizing $\Sigma_{i,\mathcal{I}'}$ (the set of inputs actually presented for that computational step) from $\Sigma_{\mathcal{I}}$ (the set of possible inputs for the total computation);

(ii.) recognizing $\Sigma_{i,\mathcal{O}'}$ (the set of outputs actually produced by the computational step) from $\Sigma_{\mathcal{O}}$ (the set of possible outputs for the total computation);

(iii.) mapping $c_i : \Sigma_{i,\mathcal{I}'} \to \Sigma_{i,\mathcal{O}'}$ (its computation for that step), $c_i \in \mathcal{C}$, the set of all computations.

Comment: This definition is a bit unusual in that it explicitly includes the recognition of the actual inputs and outputs from among the sets of possible ones, rather than just the mapping among them. The reason is that one can then place discrete, *Turing* computations firmly in a framework which accommodates both them and the analog computations of molecular machines, for example, for DNA computers. See *deterministic, nondeterministic computations*.

concerted process Two or more *primitive changes* are said to be *concerted* (or to constitute a *concerted process*) if they occur within the same *elementary reaction*. Such changes will normally (though perhaps not inevitably) be "energetically coupled." (In the present context the term "energetically coupled" means that the simultaneous progress of the *primitive changes* involves a *transition state* of lower energy than that for their successive occurrence.) In a concerted process the *primitive changes* may be *synchronous* or asynchronous.

configuration space The *configuration space* of a mechanical system is an n dimensional *manifold* Q with local (position) coordinates $q^i, \ldots, q^n$. Then T^*Q is the *momentum phase space*.

conformal mapping A map which preserves angles. A *conformal diffeomorphism* on a *Riemannian manifold* (M, g) is a diffeomorphism $\phi : (M, g) \to (M, g)$ such that $\phi^* g = f^2 g$ for a nowhere vanishing function f.

conformation The spatial arrangement of the atoms affording distinction between stereoisomers which can be interconverted by rotations about formally single bonds. Some authorities extend the term to include inversion at trigonal pyramidal centers and other *polytopal rearrangements*.

See also *tub conformation*.

conjugate momenta For a *Lagrangian* $L(q_i, \dot{q}_i)$ the *conjugate momenta* p_i are

$$p_i = \frac{\partial L}{\partial \dot{q}^i}$$

conjugated system (conjugation) In the original meaning a *conjugated system* is a *molecular entity* whose structure may be represented as a system of alternating single and multiple bonds. For example, $CH_2 = CH - CH = CH_2$, $CH_2 = CH - C \equiv N$. In such systems, conjugation is the interaction of one p-orbital with another across an intervening σ-bond in such structures. (In appropriate *molecular entities* d-orbitals may be involved.) The term is also extended to the analogous interaction involving a p-orbital containing an unshared electron pair, e.g., $Cl - CH = CH_2$.

connected Describing a *topological space* $(X, \tau(X))$ in which the empty set, $\emptyset$, and the whole space, X, are the only subsets which are both closed and open. The open interval $I = (0, 1) \subset \mathbb{R}$ is connected in the standard topology. The union of I and the open interval $(-1, 0)$ is not connected. In fact, both I and $(-1, 0)$ are at the same time open and closed, since they complement each other.

See *pathwise connected*.

connected graph A graph $\mathsf{G}(\mathcal{V}, \mathcal{E})$ is *connected* if there is a path between any two vertices, $\{v_i, v_j\} \in \mathcal{V}$, for all pairs of vertices.

Comment: Visually, *connected graphs* are those which are in "one piece." Note that if a *connected graph* contains a cycle, breaking the cycle once is guaranteed to preserve the connectedness of the graph. The effect of subsequent breaks will depend on the topology of the graph.

connection of a bundle A *distribution* (of rank $m = \dim(M)$) of planes $\Delta_b \subset T_b B$ over the total space B of a *bundle* $(B, M, \pi; F)$. The spaces $\Delta_b \subset T_b B$ are required to be nonvertical so that $\Delta_b \oplus V_b \simeq T_b B$, where V_b denotes the subspace of vertical vectors.

connectivity In a chemical context, the information content of a *line formula*, but omitting any indication of *bond* multiplicity.

conservation law A *conservation law* or *conserved quantity* of a *vector field* X is any function F such that $F \circ \phi_t = F$ where ϕ_t is the *flow* of X. See *constant of motion*.

conservative vector field A *vector field* F on a region in $\mathbb{R}^n$ is called *conservative* if it is the gradient of a function f, i.e., $F = \nabla f$.

conserved current For a *time-independent Lagrangian system* a *conserved current* is a *first integral*.

In field theory, it is a $(m - 1)$-form $\mathcal{E}$ over some jet-prolongation $J^h B$ such that it is closed once evaluated on a critical section σ, i.e., such that

$$\mathrm{d}[(j^h \sigma)^* \mathcal{E}] = 0$$

See *Lagrangian system*.

constant of motion A function $f : M \to \mathbb{R}$ on a *manifold* M is called *a constant of motion* of a *vector field* X on M if $f \circ F_t = f$, where F_t is the *flow* of X. If X_H is a *Hamiltonian vector field*, then f is a constant of motion of X_H if the *Poisson bracket* $\{f, H\} = 0$. See *conservation law*.

constitutional unit An atom or group of atoms (with pendant atoms or groups, if any) comprising a part of the essential structure of a *macromolecule*, an *oligomer molecule*, a *block*, or a *chain*.

contact form A 1-form over $J^k B$ which identically vanishes on holonomic sections $j^k \sigma$. They are locally expressed as

$$\begin{cases} \omega^i = \mathrm{d}y^i - y_\nu^i \mathrm{d}x^\nu \\ \omega_\mu^i = \mathrm{d}y^i - y_{\mu\nu}^i \mathrm{d}x^\nu \\ \cdots \\ \omega_{\mu_1 \ldots \mu_{k-1}}^i = \mathrm{d}y^i - y_{\mu_1 \ldots \mu_{k-1} \nu}^i \mathrm{d}x^\nu. \end{cases}$$

One defines three different *ideals* of contact forms:

$\Omega_c(J^k B)$ is the bilateral *ideal* generated algebraically by contact 1-forms $(\omega^i, \omega_\mu^i, \ldots, \omega_{\mu_1 \ldots \mu_{k-1}}^i)$.

$\Omega_c^*(J^k B)$ is the differential bilateral *ideal* generated by contact 1-forms (i.e., it includes products and exterior differentials of generators). Accordingly, this includes also the 2-forms $d\omega^j_{\mu_1 \dots \mu_{k-1}} = -dy^j_{\mu_1 \dots \mu_{k-1}\nu} \wedge dx^\nu$.

$\bar{\Omega}^*_c(J^k B)$ is the *ideal* of all forms vanishing on all holonomic sections, including all $(m+h)$-forms with $h > 0$.

continuity equation   The law of conservation of mass of a fluid with density $\rho(x, t)$ and velocity v is called *continuity equation*

$$\frac{\partial \rho}{\partial t} + div\,(\rho v) = 0.$$

continuous function   A function between *topological spaces* such that the inverse image of every open set is open.

continuous spectrum   The *continuous spectrum* $\sigma_c(T)$ of a linear operator T on a *Hilbert space* **H** consists of all $\lambda \in C$ such that $T - \lambda I$ is a one-to-one mapping of **H** onto a *dense* proper subspace of **H**.

continuously differentiable   A function $f : U \subset \mathbb{R}^n \to \mathbb{R}^m$ whose partial derivatives $\frac{\partial f_i}{\partial x_j}$ exist and are continuous is called *continuously differentiable* or of class C^1.

contractible   A topological space X is *contractible* when the identity map id : $M \to M$ (defined by id$(x) = x$)) and the *constant map* $c_x : M \to M$ (defined by $c_x(y) = x$)) are *homotopic*. See *homotopy*.

contraction   (1) (in metric topology) A map $T : (M, \rho) \to (M, \rho)$ of a metric space (M, ρ) into itself such that there exists a constant $\lambda, 0 < \lambda < 1$ such that

$$\rho(Tx, Ty) \leq \lambda \rho(x, y), for\ all\ x, y \in M.$$

(2) (in networks) In a network (or graph), the replacement of a larger subnetwork by a smaller using a sequence of mathematical operations. Network N′ is a *contraction* of network N if N′ can be obtained from N by a sequence of node, edge, parameter, or label combination operations.

Comment: The nature of the combination operations and their operands can vary considerably depending on the intended result of the contraction. However, no other nonmathematical information is required for contractions of models of biological systems, such as biochemical networks (unlike *expansion*). Such information may be beneficial, for example, in suggesting appropriate approximations. An *expansion* is the reverse operation.

contraction mapping theorem   Let $T : (M, \rho) \to (M, \rho)$ be a contraction mapping of a *complete metric space* (M, ρ) into itself. Then T has a unique fixed point; i.e., there exists a unique $x_0 \in M$ such that $Tx_0 = x_0$.

contravariant/covariant tensor   A tensor T on a vector space E, *contravariant* of order r and *covariant* of order s is a multilinear map

$$T : E^* \times \cdots \times E^* \times E \times \cdots \times E \to \mathbb{R}$$

(r copies of E^* and s copies of E).

convergent sequence   A sequence $\{x_n\}$ having a limit L. That is, for every neighborhood U of L, we have $x_n \in U$ for all except finitely many n.

convex combinations   See *convex hull*.

convex hull   The *convex hull* of a set A in a *vector space* X is the set of all *convex combinations* of elements of A, i.e., the set of all sums $t_1 x_1 + \cdots t_n x_n$, with $x_i \in A, t_i \geq 0, \sum t_i = 1, n$ arbitrary.

convex set   A set $C \subset X$ is *convex* if for any $x, y \in C, tx + (1-t)y \in C$ for all t $(0 < t < 1)$.

convolution   The *convolution* $f * g$ of two functions f and g is given by

$$f * g(x) = \int f(x - y)g(y)dy.$$

coordinate   See *basis of a vector space*.

coordinate patch   See *chart*.

coordinate transition function   See *chart*.

coordination The formation of a covalent *bond*, the two shared electrons of which have come from only one of the two parts of the *molecular* entity linked by [iut,] as in the reaction of a *Lewis acid* and a *Lewis base* to form a *Lewis adduct*; alternatively, the bonding formed in this way. In the former sense, it is the reverse of *unimolecular heterolysis*. "Coordinate covalence" and "coordinate link" are synonymous (obsolescent) terms. The synonym "dative bond" is also obsolete. (The origin of the bonding electrons has by itself no bearing on the character of the bond formed. Thus, the formation of methyl chloride from a [methylu cation] and a chloride ion involves coordination; the resultant bond obviously differs in no way from the C–Cl bond in methyl chloride formed by any other paty, e.g., by colligation of a methyl radical and a chlorine atom.)

The term is also used to describe the number of ligands around a central atom without necessarily implying two-electron bonds.

copolymer A *polymer* derived from more than one species of *monomer*.

Note: Copolymers that are obtained by *copolymerization* of two *monomer* species are sometimes termed bipolymers, those obtained from three *monomers* terpolymers, those obtained from four *monomers* quaterpolymers, etc.

copolymerization *Polymerization* in which a *copolymer* is formed.

corrosion An irreversible interfacial reaction of a material (metal, ceramic, polymer) with its environment which results in consumption of the material or in dissolution into the material of a component of the environment.

Often, but not necessarily, corrosion results in effects detrimental to the usage of the material considered. Exclusively physical or mechanical processes such as melting or evaporation, abrasion or mechanical fracture are not included in the term corrosion.

cosecant The function $\csc(x) = \frac{1}{\sin(x)}$. See *sine*.

cosine The function $\cos(x) = \frac{e^{ix}+e^{-ix}}{2}$. Geometrically, it is the ratio of the lengths of adjacent side to hypotenuse for a right triangle with angle x for $0 < x < \frac{\pi}{2}$.

cosmological constant In general relativity it is possible to add the term $\Lambda g^{\mu\nu}$ to the *energy-momentum tensor*, where Λ is the *cosmological constant*, and $g^{\mu\nu}$ is the space-time metric tensor.

cotangent The function $\cot(x) = \frac{\cos x}{\sin x}$. See *cosine, sine*.

Coulomb gauge In *gauge theory* the *Coulomb gauge* is defined by taking $\partial^i A_i = 0$, where A is the vector potential.

Coulomb potential The *Coulomb potential* $\Phi(x)$ of a source function f on $\mathbb{R}^n$ is given by

$$\Phi(x) = \int_{\mathbb{R}^n} \|x - y\|^{2-n} f(y) dy.$$

For $n = 3$, this is the potential function for Newton's theory of gravitation.

countable additivity A *measure* μ has the property of *countable additivity* if given $A_1, A_2, \ldots$ a sequence of pairwise disjoint *measurable* sets then

$$\mu(\bigcup_{i=1}^{\infty} A_i) = \sum_{i=1}^{\infty} \mu(A_i).$$

countable set A set is *countable* if it is either finite or denumerable. See also *cardinality, denumerably infinite set, finite set, infinite set,* and *uncountably infinite set*.

covalent bond A region of relatively high electron density between nuclei which arises at least partly from sharing of electrons and gives rise to an attractive force and characteristic internuclear distance.

covariant derivative On a *Riemannian manifold* (M, g) the *covariant derivative* of a *vector field* Y along the *vector* field X is the *vector* field $\nabla_X Y$ locally given by

$$\nabla_X Y = X^j Y^k \Gamma^i_{jk} \frac{\partial}{\partial q^i} + X^j \frac{\partial Y^k}{\partial q^j} \frac{\partial}{\partial q^k}$$

where Γ^i_{jk} are the *Christoffel symbols* of g.

The *covariant derivative* of a section σ of a *bundle* $(B, M, \pi; F)$ with respect to a connection Γ is defined by

$$\nabla_X \sigma = T\sigma(X) - \Gamma(X)$$
$$= X^\mu(\partial_\mu \sigma^i - \Gamma_\mu^i(x, \sigma))\partial_i$$

where $X = X^\mu \partial_\mu$ is a *vector* field over M, T is the tangent map and $\Gamma(X)$ is the *horizontal lift* of X with respect to the connection Γ. It is a *vertical vector field* defined over the section σ, i.e., a section of the *bundle* $TB \to M$ which projects over the section $\sigma : M \to B$.

covariant tensor   See *contravariant tensor*.

covering space of M   A space C and a projection $\pi : C \to M$ which is a *local diffeomorphism*. A covering space can be regarded as a *bundle* with a discrete standard fiber. *Covering spaces* are characterized by the property of the lifting of curves: if γ is a curve in M based at the point $x \in M$ and $b \in B$ is a point projecting on $x = \pi(b)$, then there exists a unique curve $\hat{\gamma}$ in B based at b which projects over $\gamma(t) = \pi(\hat{\gamma}(t))$. Furthermore, the lift of two *homotopic* curves produces two curves of B which are still homotopic.

critical configuration   See *Hamilton principle*.

critical point   A *critical point* of a *vector field* X is a point x_0 such that $X(x_0) = 0$.

crystallographic group   Crystals are formed with symmetries. In three-dimensional space, these symmetries are represented by the *crystallographic groups*. These include rotation, reflection (point group), and translation (lattice symmetry) as basic elements, and their composition (e.g., screw axes and glide planes) lead to a finite group. More specifically, 32 point groups combined with 14 Bravais lattices (7 primitive and 7 nonprimitive) result in 230 unique space groups which describe the only ways in which identical objects may be arranged in an infinite 3D lattice. A simpler, practical approach leading to the same result studies translations in a lattice, introducing two new kinds of symmetry operations, screw axes and glide planes. (*cf.* G.H. Stout and L.H. Jensen, *X-Ray Structure Determination: A Practical Guide*, 2nd ed., John Wiley & Sons, New York, 1989).

curl   The *curl* of a *vector* field $F = (F_1, F_2, F_3)$ on $\mathbb{R}^3$ is

$$curl\ F = \nabla \times F$$
$$= \left(\frac{\partial F_3}{\partial y} - \frac{\partial F_2}{\partial z}\right)\mathbf{i} + \left(\frac{\partial F_1}{\partial z} - \frac{\partial F_3}{\partial x}\right)\mathbf{j}$$
$$+ \left(\frac{\partial F_2}{\partial x} - \frac{\partial F_1}{\partial y}\right)\mathbf{k}$$

curve (in a topological space X)   A continuous map $\gamma : \mathbb{R} \to X$. Sometimes the domain is restricted to an interval $I \subset \mathbb{R}$; if the origin $0 \in R$ is a point of I, then the *curve* is said to be *based at* $x = \gamma(0)$. If X is a *differentiable manifold*, the curve γ is usually required to be differentiable.

Not to be confused with the *trajectory* or *path* which is the image of the *curve*, i.e., the subset $\Im(\gamma) \subset X$ of the space X.

One can suitably define the *composition* of two curves $\gamma, \lambda : [0, 1] \to X$ provided that $\gamma(1) = \lambda(0)$. Notice that even if the two curves γ and λ are differentiable, the composition $\lambda * \gamma$ can be nondifferentiable.

cycle   A path through the graph $\mathsf{G}(\mathcal{V}, \mathcal{E})$ consisting of nodes $\mathcal{V}' = \{v_1, v_2, v_3, \ldots, v_n\}$ and edges $\mathcal{E}' = \{(v_1, v_2), (v_2, v_3), \ldots, (v_n, v_1)\}$, $\mathcal{V}' \subseteq \mathcal{V}$, $\mathcal{E}' \subseteq \mathcal{E}$. The *length* of the cycle is n.

Comment: It is the property of "pathness," that there is a sequence of edges connecting the nodes in the cycle, which distinguishes a *cycle* from a disconnected subgraph. This is the only part of the definition which might otherwise not be intuitive.

cytoskeleton   A dynamic *polymer* network underneath the cell membrane. It is mostly responsible for the mechanics, i.e., shape and motility, of cells. The main types of filaments in the network are actin filaments, microtubules, and intermediate filaments. There are many other different proteins involved in the network. They regulate the state and dynamics of the network.

D

dalton   The unit of molecular weight, in g/mol.

Comment: A mole (abbreviated mol) is an Avogadro's number ($\mathcal{N}$) of molecules, approximately equal to 6.02252×10^{23} molecules/mol.

Darboux's theorem   On any *symplectic manifold* (M, ω) there exist local coordinates (*canonical coordinates*) $(x^1, \ldots, x^n, y^1, \ldots, y^n)$ in which the symplectic form ω takes the form

$$\omega = \sum_{i=1}^{n} dx^i \wedge dy^i .$$

dark current   See *responsivity*.

dark output   See *responsivity*.

dark resistance   See *responsivity*.

data model   A computational method for implementing a *domain model*.

Comment: This is the database management system, object-oriented model, set of MatLab functions, etc., which [reify] the ideas in the domain model. Thus, it is the computational machinery, not (per se) the intellectual basis of the abstract model of the domain.

decreasing function   A function $f(x)$ such that $f(y) \leq f(x)$ whenever $x \leq y$.

definite integral   Let f be a function defined on a closed interval $[a, b]$ and P a partition $a = x_0 < x_1 < x_2 < \cdots < x_n = b$ of $[a, b]$. Let $|P|$ denote the length of the longest subinterval in the partition P and let $\bar{x}_i \in [x_{i-1}, x_i]$ be arbitrary. If the limit

$$\lim_{|P| \to 0} \sum_{i=1}^{n} f(\bar{x}_i)(x_i - x_{i-1})$$

exists, f is called (Riemann) integrable and

$$\int_a^b f(x)dx = \lim_{|P| \to 0} \sum_{i=1}^{n} f(\bar{x}_i)(x_i - x_{i-1})$$

is called the *definite integral* (or Riemann integral) of f from a to b.

degree   The number of edges incident to a node; here notated as $d(v)$.

Comment: The number of reactions in which a compound participates is its *degree*. Similarly, the number of compounds participating in a reaction is the degree of the reaction. Since the best-known biochemical network described has an *enzyme* for every reaction, one usually means *nonenzymatic degree*, when applying the word to reactions. Note this would not be the case for other versions of the biochemical network which include spontaneous reactions, or detailed kinetic and mechanistic views of reactions in which the enzyme does not necessarily appear in the reaction equations as a formal catalyst.

degree of polymerization   The number of *monomeric units* in a *macromolecule* or *oligomer molecule*, a *block*, or a *chain*.

degrees of freedom   A system of N particles, free from constraints, has $3N$ independent coordinates (positions and momenta) called *degrees of freedom*.

delocalization   A quantum mechanical concept most usually applied in organic chemistry to describe the π-bonding in a *conjugated system*. This bounding is not localized between two atoms: instead, each link has a "fractional double bond character" or *bond order*. There is a corresponding "delocalization energy," identifiable with the stabilization of the system compared with a hypothetical alternative in which formal (localized) single and double bonds are present. Some degree of delocalization is always present and can be estimated by quantum mechanical calculations.

delta function   See *Dirac delta function*.

dense   A subset S in a space X is *dense* (in X) if its closure $\bar{S} = X$.

denser subgraph   A subgraph G' of the biochemical network such that the degree of each reaction node $v_{r,i}$ is greater than or equal to x, and the degree of each compound node $v_{m,j}$ is less than or equal to y; that is, $d(v_{r,i}) \geq x$, $d(v_{m,j}) \leq y$.

Comment: Because any two denser subgraphs formed with the same values of x and y are not necessarily *isomorphic*, denser subgraphs are not necessarily motives. Familiarly known as a *blob*.

denumerably infinite set   A set which is equivalent to the set of natural numbers, $\mathbb{N}$. (There is a *bijection* between the set and $\mathbb{N}$.) See also *cardinality*, *finite set*, *infinite set*, and *uncountably infinite set*.

deoxyribonucleic acids (DNA)   High-molecular-weight linear polymers, composed of *nucleotides* containing deoxyribose and linked by phosphodiester bonds; DNA contain the genetic information of organisms. The double-stranded form consists of a double helix of two complementary chains that run in opposite directions and are held together by hydrogen bonds between pairs of the complementary *nucleotides* and Hoogsteen (stacking) forces.

deRham cohomology group   The quotient groups of *closed forms* by *exact forms* on a *manifold* M are called the *deRham cohomology groups* of M. The kth *deRham cohomology group* of M is

$$H^k(M) = ker\ \mathbf{d}^k / range\ \mathbf{d}^{k-1},$$

where $\mathbf{d}^k$ is the exterior derivative on k forms.

derivation (of an R-algebra A)   Given R a *ring*, an R-linear map $D : A \to A$ (i.e., $D(\lambda x + \mu y) = \lambda D(x) + \mu D(y)$) such that the *Leibniz rule* holds, i.e., $\forall x, y \in A\ D(xy) = D(x)y + xD(y)$. A derivation on a *manifold* M is a derivation of the *function algebra* $A = C^\infty(M)$.

Example: let $F(\mathbb{R})$ be the $\mathbb{R}$-algebra of real valued (differentiable) functions $f : \mathbb{R} \to \mathbb{R}$ over $\mathbb{R}$. The ordinary *derivative* $D : F(\mathbb{R}) \to F(\mathbb{R})$ is a *derivation*.

Notice that A is not required to be an associative algebra. In particular the definition applies to *Lie algebras* where the Leibniz rule reads as $D([x, y]) = [D(x), y] + [x, D(y)]$.

Example: let L be a *Lie algebra*. For all $x \in L$, the map $\text{ad}_x : L \to L$ defined by $\text{ad}_x : y \mapsto [x, y]$ is a *derivation*.

derivative   The *derivative* of a *function* $f : \mathbb{R} \to \mathbb{R}$ at a point x is the *function* f' defined by

$$f'(x) = \lim_{h \to 0} \frac{f(x + h) - f(x)}{h}$$

provided that this limit exists, f is called *differentiable* at the point x with derivative $f'(x)$. More generally, if X, Y are *Banach spaces* $U \subset X$ open, and $f : U \subset X \to Y$, then the *Frechet derivative* of f is map $Df : U \to L(X, Y)$, where $L(X, Y)$ is the vector space of *bounded linear operators* for X to Y defined by

$$Df(x)h = \lim_{t \to 0} \frac{f(x + th) - f(x)}{t}.$$

If this limit exists f is called *differentiable* at x with (total or *Frechet*) derivative $Df(x)$.

deterministic computation   A *deterministic computation* $\mathcal{C}_d$ specifies a computation $\mathcal{C} : \Sigma_{\underline{\mathcal{I}}'} \to \Sigma_{\mathcal{O}'}$ such that the cardinalities (denoted $\bar{S}$, where S is any set) of the sets of presented inputs and produced outputs ($\Sigma_{\mathcal{I}'}$ and $\Sigma_{\mathcal{O}'}$, respectively) are 1 ($\overline{\Sigma}_{\mathcal{I}'} = \overline{\Sigma}_{\mathcal{O}'} = 1$); each computational step is a one-to-one mapping between the presented input and the produced output ($\forall c_i$, $c_i \in \mathcal{C}_d$ is one-to-one); and the probability that each symbol in the sets of presented inputs and produced outputs exists is 1 ($\forall \sigma_{i,\mathcal{I}'}, \sigma_{i,\mathcal{I}'} \in \Sigma_{\mathcal{I}'}$, $\forall \sigma_{i,\mathcal{O}'}, \sigma_{i,\mathcal{O}'} \in \Sigma_{\mathcal{O}'}, P_e(\sigma_{i,\mathcal{I}'}) = P_e(\sigma_{i,\mathcal{O}'}) = 1$).

Comment: As with the definitions of *computation*, *nondeterministic*, and *stochastic computations*, the goal here is to place the usual theory of computing within a framework that accommodates molecular computers. Like any other system of chemical reactions, a molecular computer is surrounded by many copies of its possible inputs and outputs. These copies of symbols are populations of symbols, each type of symbol occuring at some frequency in its respective population (P_e). $P_e(\sigma_i)$ for any particular σ_i can vary depending on the constitution of $\Sigma_{\mathcal{I}}, \Sigma_{\mathcal{O}}$, and the properties of a particular c_i; so there exists a probability density function over Σ. For computation over a number of steps, a *vector* $\mathbf{P}_e$ of probabilities for each step to each σ_i would be assigned. Notice that a unit cardinality does not imply that the length of the input and output tokens is one. See also *nondeterministic* and *stochastic computations*.

device   A subnetwork of the biochemical network with distinct dynamical (and perhaps biochemical) properties.

dextralateral   The set of obligatorily co-reacting species arbitrarily written on the right-hand side of a formal reaction equation.

Comment: Formal reaction equations represent two sets of molecular species, one written on the left and the other on the right-hand side of the equation. The placement of a set on a side of the equation is completely arbitrary. The equation is understood to be symmetric in that the reaction's chemistry proceeds in both the "forward" (left to right) and "backward" (right to left) simultaneously, until *dynamic equilibrium* is achieved. For the (bio)chemistry to proceed, each member of a set of coreacting species must be present. For each, the higher its concentration, the easier it is to observe the reaction and the faster the reaction will occur.

Since the reactions are reversible, the equations are symmetric, and the placement of sets of coreacting species in the equation is arbitrary, the sets of species are designated *sinistralateral* and *dextralateral*. These designations distinguish them from the sets of *substrates* and *products*: these terms indicate the role a molecule plays in the reaction. See also *direction, dynamic equilibrium, formal reaction equation, microscopic reversibility, product, rate constant, reversibility, sinistralateral,* and *substrate*.

diagonalizable   A linear transformation $T : V \to V$ on a *vector space* V is called *diagonalizable* (or semi-simple) if there exists a *basis* of V consisting entirely of *eigenvectors* of T. See *linear*.

diagram level (in x-ray spectroscopy)   A level described by the removal of one electron from the configuration of the neutral ground state. These levels form a spectrum similar to that of a one-electron or hydrogen-like atom but, being single-valency levels, have the energy scale reversed relative to that of single-electron levels. Diagram levels may be divided into valence levels and core levels according to the nature of the electron vacancy. Diagram levels with orbital angular momentum different from zero occur in pairs and form spin doublets.

diagram line (in x-ray spectroscopy)   See *characteristic x-ray emission* and *x-ray satellite*.

diameter   For any bounded domain $K \subset \mathbb{R}^n$ we define its diameter by

$$\text{diam}(K) := \sup\{|\mathbf{x} - \mathbf{y}|, \mathbf{x}, \mathbf{y} \in K\}.$$

diffeomorphism   A *diffeomorphism of class* C^k is a C^k differentiable map f which is invertible such that f^{-1} is also of class C^k.

differentiable   See *derivative*.

differentiable manifold   A topological space M is called *differentiable manifold of class* C^p if each point $x \in M$ has a *neighborhood* $U(x) \subset M$ (called local chart or coordinate patch) which is *homeomorphic* to an open set in a *vector space* ($\mathbb{R}^n$), and such that the change of coordinates (coordinate transition functions) are differentiable maps of class C^p. See *chart*.

differential equation   An equation involving a *function* and some of its *derivatives*, e.g., $f''(x) + f(x) = 0$.

differential form   A *differential form of degree* r (or an r-form) on an open set U in a *vector space* X is a *smooth map* $\omega : U \to \wedge^r X^*$ from U into the rth alternating product of X^*.

differential operator   Let E and F be *vector bundles* over a *manifold* M and $C^\infty(E)$, $C^\infty(F)$ the spaces of *smooth* sections. A *differential operator* is a *linear* map $L : C^\infty(E) \to C^\infty(F)$ such that $L(fg) = (Lf)g + (Lg)f$ for all $f, g \in C^\infty(E)$.

diffusion equation   The *diffusion equation* or *heat equation* is of the following form, for $u = u(x, t), x \in \mathbb{R}^n$

$$(\partial_t - \Delta)u = 0 \,.$$

e.g., in two dimensions $\frac{\partial^2 u(x,t)}{\partial x^2} = \frac{\partial u(x,t)}{\partial t}$.

diffusion-driven instability   This is a bifurcation phenomenon in nonlinear diffusion-reaction equations. If when all the diffusion terms are absent, the remaining nonlinear ODE has a stable fixed point, but when a diffusion term is present, the spatially homogeneous solution corresponding to the stable fixed point is unstable, we say the system has a *diffusion-driven instability*.

Dini's theorem   Let $\{f_n\}$ be a sequence of *continuous functions* converging pointwise to a *continuous function* f. If $\{f_n\}$ is monotone increasing sequence, then the convergence is uniform.

dipole–dipole interaction   Intermolecular or intramolecular interaction between molecules or groups having a permanent electric dipole moment. The strength of the interaction depends on the distance and relative orientation of the dipoles. The term applies also to intramolecular interactions between bonds having permanent dipole moments.

Dirac   Paul Adrien Maurice Dirac (1902–1984), Swiss/English theoretical physicist, founder of relativistic quantum mechanics. Nobel prize for physics 1933 (shared with Schrödinger).

Dirac delta function   The *generalized function* $\delta(x - x_0)$ defined by

$$\int_{-\infty}^{\infty} f(x)\delta(x - x_0)dx = f(x_0).$$

Dirac delta measure   The *measure* δ_y located at some arbitrary, but fixed $y \in \mathbb{R}^n$ is defined for $A \subset \mathbb{R}^n$ as

$$\delta_y(A) = \begin{cases} 1 \text{ if } y \in A \\ 0 \text{ if } y \notin A. \end{cases}$$

Dirac equation   The equation

$$(i \not{\partial} - m)\psi = 0$$

where ψ is a *wave function* describing a relativistic particle of mass m and $\not{\partial} := \gamma^\mu \delta_\mu$ with γ^μ the *Dirac gamma matrices*.

Dirac gamma matrices   The 4×4 matrices $\gamma^0 = \begin{pmatrix} I & 0 \\ 0 & -I \end{pmatrix}$, $\gamma^j = \begin{pmatrix} 0 & \sigma^j \\ -\sigma^j & 0 \end{pmatrix}$, $j = 1, 2, 3$, where I is the 2×2 identity matrix and σ^j are the Pauli matrices $\sigma^1 = \begin{pmatrix} 0 & 1 \\ 1 & 0 \end{pmatrix}$, $\sigma^2 = \begin{pmatrix} 0 & -i \\ i & 0 \end{pmatrix}$, $\sigma^1 = \begin{pmatrix} 1 & 0 \\ 0 & -1 \end{pmatrix}$, $i^2 = -1$.

Dirac Laplacian   The square root of the *Dirac operator*, i.e., we have $/D^2 \psi = \Box \psi$, the d'Alembert operator.

Dirac operator   A linear first-order partial *differential operator* with nonconstant coefficients, defined between sections of the spin bundle. It is defined as *covariant derivative* followed by *Clifford product*. Locally the *Dirac operator* is given by

$$\not{D}\psi = \sum_\mu \gamma^\mu \frac{\delta}{\delta x^m u} \psi$$

where γ^μ are the *Dirac gamma matrices*.

Dirac spinors   Elements in the full complex spin representation as opposed to elements in the half spin representation which are called *Weyl spinors* and elements in the real spin representation which are called *Majorana spinors*.

directed edge   A sequence of two nodes (ordered pair) in a graph; often denoted $\langle v_i, v_{i+1} \rangle$.

Comment: These are the edges that are drawn with arrows, indicating a direction of travel or flow. See *edge*.

direction   In a *formal reaction equation*, the consumption of *sinistralateral* coreactants (the "forward" direction); equally, the consumption of *dextralateral* coreactants (the "reverse" direction).

Comment: Chemical and biochemical reactions can be thought of as a composite of two reactions. One that consumes sinistralateral coreactants, producing dextralateral ones or proceeding left to right and the opposite. These two directions are often represented as half reactions which are formal equations that specify only one of the two directions. The intrinsic rates of the two directions, as measured by their respective *rate constants* can differ very significantly. See also *dextralateral, dynamic equilibrium, formal reaction equation, microscopic reversibility, product, rate constant, reversibility, sinistralateral,* and *substrate*.

Dirichlet boundary condition   Specification of the value of the solution to a *partial differential equation* along a bounding surface.

Dirichlet problem   The problem of finding solutions of the *Laplace equation* $\Delta u = 0$ ($u_{xx} + u_{yy} = 0$ in two dimensions) that take on given boundary values.

discrete spectrum   The *discrete spectrum* or *point spectrum* $\sigma_p(A)$ of a *linear operator* A is the set of all λ for which $(\lambda I - A)$ is not one-to-one, i.e., $\sigma_p(A)$ is the set of all *eigenvalues* of A.

discrete topology   In the *discrete topology* on a set S every subset of S is an open set.

disjoint   Two collections, especially sets such as A and B, are said to be *disjoint* if $A \cap B = \emptyset$.

dissipative   A *linear operator* $T : H \to H$ on a *Hilbert space* $(H, < \ , \ >)$ is called *dissipative* if $Re < Tu, u > \leq 0$ for all $u \in H$.

dissociation   (**1**) The separation of a *molecular entity* into two or more *molecular entities* (or any similar separation within a polyatomic molecular entity). Examples include *unimolecular heterolysis* and *homolysis*, and the separation of the constituents of an *ion pair* into free ions. (**2**) The separation of the constituents of any aggregate of *molecular entities*.

In both senses dissociation is the reverse of *association*.

distribution   A *distribution* on an open set $U \subset \mathbb{R}^n$ is a continuous linear functional on $C_c^\infty(U)$ (*smooth* functions with compact support).

distribution (of subspaces)   A family of subspaces $\Delta_x \subset T_x M$ one for each $x \in M$. When the dimension of the subspaces Δ_x is constant with respect to x such a dimension k is called the *rank* of the distribution. One should add some regularity condition in order not to allow too odd dependence of the subspace on the point x. Usually one asks that locally there exist k local vector fields, called *generators of* Δ, spanning Δ_x at each point of an open set $U \subset M$.

divergence   Let M be a *manifold* with volume Ω and X a vector field on M. Then the unique function $div_\Omega X \in C^\infty(M)$ such that the *Lie derivative* $L_X \Omega = (div_\Omega X)\Omega$ is called the *divergence* of X. If $M = \mathbb{R}^3$ and $X = (f_1, f_2, f_3)$ then

$$div X = \frac{\partial f_1}{\partial x} + \frac{\partial f_2}{\partial y} + \frac{\partial f_3}{\partial z}.$$

divergence theorem   Also called *Gauss's theorem*. Let Ω be a region in $\mathbb{R}^3$ and $\partial\Omega$ the oriented surface that bounds Ω, and denote by n the unit outward normal vector to $\partial\Omega$. Let X be a vector field defined on Ω. Then

$$\int_\Omega (div\ X)dV = \int_{\partial\Omega} (X \cdot n)dS.$$

divergent sequence   A sequence that is not convergent.

dividing surface   A surface, usually taken to be a hyperplane, constructed at right angles to the *minimum-energy path* on a *potential-energy surface*. In conventional *transition-state theory* it passes through the highest point on the minimum-energy path. In generalized versions of transition-state theory the dividing surface can be at other positions; in *variational transition-state theory* the position of the dividing surface is varied so as to get a better estimate of the rate constant.

DNA   See *deoxyribonucleic acids*.

DNA supercoil   A DNA molecule has a double helical structure (see *double helix*). Consider the axis of the double helix as a space curve; it is known experimentally that the *curve* can have non-planar geometry. For example, it can be solenoidal itself, thus the name supercoil. DNA from some organisms have their helical axis being a closed space curve with topological linking numbers, i.e., knot. (*cf.*, W.R. Bauer, F.H.C. Crick, and J.H. White, *Sci. Am.*, 243, 118, 1980).

domain   In computer science, a *domain* is a discipline, an area of physical reality, or a thought modeled by a representation; in essence, its subject. In mathematics, a *domain* is the set on whose members a relation operates.

Comment: For further comments on the mathematical sense of *domain*, see *relation*. See also *image* and *range*.

domain model   A formal model of a particular domain (in the computational sense).

Comment: It is this that a database or artificial intelligence system, or indeed any abstract model of a phenomenon, reifies. It is distinguished from a *data model*, which is an implementation method (such as relational, object-oriented, or declarative database).

dominated convergence theorem   Let $\{f_n\}$ be a sequence in L^1 such that $f_n \to f$ almost everywhere and there exists a nonnegative $g \in L^1$ such that $|f_n| \leq g$ almost everywhere for all n. Then $f \in L^1$ and $\int f = \lim_{n \to \infty} \int f_n$.

donor   A compound which breaks a chemical bond, yielding a substituent group which forms a new bond in a bimolecular chemical or *biochemical reaction*. See *acceptor*.

dot product   See *angle between vectors*.

double helix   The structure of DNA in all biological species is in the form of a double helix made of two chain molecules. Each chain is a *polymer* made of four types of nucleotide: A (adenine), G (guanine), T (thymine), and C (cytosine). The structure that was first proposed by J.D. Watson and F.H.C. Crick immediately leads to a possible mechanism of biological heredity. This was later confirmed by experiments and hence provides a molecular basis for genetics.

drawing   See *rendering*.

dual   See *dual vector space, complement*.

dual basis   Let E be a finite dimensional *vector space* with basis $(e_1, \ldots, e_n)$. The *dual basis* $(\alpha^1, \ldots, \alpha^n)$ of the dual space E^* is defined by $\alpha^j(e_i) = \delta_i^j$, where $\delta_i^j = 1$ if $j = i$ and 0 otherwise.

dual vector space   Let V be a *vector space* (over the field $\mathbb{K}$); the set V^* of all linear functionals $\alpha : V \to \mathbb{K}$ can be endowed with a structure of *vector space* by defining $(\lambda \alpha + \mu \beta)(v) = \lambda \, \alpha(v) + \mu \, \beta(v)$ (where $\lambda, \mu \in \mathbb{K}; v \in V$). The *vector space* V^* so obtained is called the *(algebraic) dual vector space of V*.

If V is a topological *vector space* the set of all linear and continuous functionals $\alpha : V \to \mathbb{K}$ with the standard linear structure introduced above is called the *topological dual space* of V, and it is still denoted by V^*.

If V is finite dimensional, then $\dim(V) = \dim(V^*)$. See also *dual basis*.

duality techniques (Aubin-Nitsche trick)
These are used to establish a priori estimates for the discretization error of finite element schemes in norms weaker than the natural norms associated with the continuous variational problem. For example, let H, V be *Hilbert spaces*, V continuously embedded in H, and $a : V \times V \to \mathbb{C}$ a continuous and *V-elliptic* sesqui-linear form. Write $u \in V, u_h \in V_h$ for the solutions of

$$a(u, v) = f(v) \quad \forall v \in V,$$

$$a(u_h, v_h) = f(v_h) \quad \forall v_h \in V_h,$$

where V_h is some closed subspace of V (a conforming finite element space) and $f \in V'$. For $\phi \in H$ denote by $g(\phi) \in V$ the solution of

$$a(v, g(\phi)) = (\phi, v)_H \quad \forall v \in V.$$

Then, choosing $\phi := u - u_h$ we obtain via *Galerkin orthogonality*

$$\|u - u_h\|_H^2 = a(u - u_h, g(u - u_h)) \leq$$

$$\|a\| \|u - u_h\|_V \inf_{v_h \in V_h} \|g(u - u_h) - v_h\|_V$$

If $g(\phi)$ possesses extra regularity beyond merely belonging to V, the second term on the right-hand side will become very small compared to $\|u - u_h\|_H$, if the resolution of the finite element space is increased. Thus, when considering families of finite element spaces, the H-norm of the discretization error may converge asymptotically faster than the V-norm.

Duffing equation   The *Duffing equation*

$$\ddot{x} + x + \epsilon x^3 = 0$$

is an example of weakly nonlinear oscillators, i.e., small perturbations of the linear oscillator $\ddot{x} + x = 0$.

Dym's equation The nonlinear *evolution equation*

$$u_t = 2(u^{-1/2})_{xxx}.$$

dynamic equilibrium In a chemical or *biochemical reaction*, the continuous reaction of *sinistralateral* and *dextralateral* sets of coreactants, such that no net change in the concentrations of each members of both sets occurs.

Comment: What determines which *direction* of a reaction forward or backward will predominate is the relative concentration of reactants forming the two sets of obligatorily coreacting species. High concentrations of one set will drive the chemistry in the direction which consumes the reactants of that set, until the two sets are in equilibrium. See also *dextralateral, direction, formal reaction equation, microscopic reversibility, product, rate constant, reversibility, sinistralateral,* and *substrate.*

dynamic viscosity, η For a laminar flow of a fluid, the ratio of the *shear stress* to the velocity gradient perpendicular to the plane of shear.

dynamical motif A conserved pattern of dynamical regimes for a reaction or group of reactions.

Comment: Notice this is distinct from a device in that the latter is not required to exhibit conservation. See also *biochemical, chemical, functional, kinetic, mechanical, phylogenetic, regulatory, thermodynamic,* and *topological motives.*

dynamical system (1) The flow F_t of a *vector field* X on a *manifold* M; i.e., $F_t : M \to M$ is a one-parameter group of diffeomorphisms, $F_{t+s} = F_t \circ F_s$, and satisfies the *differential equation*

$$\frac{d}{dt} F_t(x) = X(F_t(x)) .$$

(2) (*autonomous dynamical system*) a pair (M, X) where M is a *manifold* and X is a *vector field* over M. An *integral curve* $\gamma : \mathbb{R} \to X$ is such that

$$\dot{\gamma} = X \circ \gamma$$

where $\dot{\gamma}$ is the tangent vector to γ.

(3) (*non-autonomous dynamical system* over M) a dynamical system $(\hat{M}, \hat{X})$ over $\hat{M} = \mathbb{R} \times M$ such that $\hat{X} = \partial_t + X(t, x)$, where $X(t, x)$ is a time-dependent vector field over M.

See *equilibrium point, first integral.*

E

edge   An unordered pair of nodes in a graph, usually denoted as a *tuple* of *arity* two with the two nodes as *arguments*. Thus $(v_i, v_j) = (v_j, v_i)$ (the edge is symmetric).

Comment: Edges are usually represented as lines or arcs in renderings of a graph or network. *Incidence relation* and *undirected edge* are synonyms. See *directed edge*.

effective Lagrangian   An *effective Lagrangian* describes the behavior of a quantum field theory at large distances (low energy).

eigenspace   See *eigenvalue*.

eigenvalue   For a *linear* operator $A : X \to X$ a scalar λ is called *eigenvalue* of A if there is a nontrivial solution x to the equation $Ax = \lambda x$. Such an x is called *eigenvector* corresponding to λ. The subspace of all solutions of the equation $(A - \lambda I)x = 0$ is called the *eigenspace* of A corresponding to λ.

eigenvector   See *eigenvalue*.

eikonal equation   In geometrical optics the *Hamilton-Jacobi equation* $H(x, dS(x)) = 0$.

Einstein   Albert Einstein (1879–1955) German-American physicist. The inventor of special and general relativity theory. Nobel prize in physics 1922. Popularly regarded as a genius among geniuses, the greatest scientist in history. *Everything should be formulated as simple as possible, but not simpler.*

Einstein equations   *Einstein's field equation* of general relativity (for the vacuum) is the system of second-order partial differential equations

$$R_{\mu\nu} = 0$$

where $R_{\mu\nu}$ is the *Ricci* curvature *tensor*.

Einstein tensor   On a pseudo-*Riemannian manifold* (M, g), the *tensor* $G_{\mu\nu} = R_{\mu\nu} - \frac{1}{2}Rg_{\mu\nu}$ where $R_{\mu\nu}$ is the *Ricci tensor* and R is the *Ricci scalar* of the *Levi-Civita connection* of the metric g.

By extension, on a *manifold* M with a connection $\Gamma^\alpha_{\beta\mu}$, it is the *tensor* with the same local expression where $R_{\mu\nu}$ is now the *Ricci tensor* and R is the *Ricci scalar* of the connection fixed on M.

Because of *Bianchi identities* we have $\nabla_\mu G^\mu_{\cdot\mu} = 0$.

electric charge, Q   Integral of the *electric current* over time. The smallest electric charge found on its own is the *elementary charge*, e, the charge of a proton.

electrode potential, E   Electromotive force of a cell in which the electrode on the left is a standard hydrogen electrode and the electrode on the right is the electrode in question.

electrodynamics   Classical *electrodynamics* is described by *Maxwell's equations* of an *electromagnetic field*. *Quantum electrodynamics (QED)* is the theory of interaction of light with matter and is described by the *Dirac equations* $\gamma_\mu(i\nabla_\mu - eA_\mu)\psi = m\psi$.

electromagnetic field   In modern geometric terms a 2-form on space-time M given as follows. Let A be a connection 1-form (vector potential). Its curvature 2-form F_A, given by $F_A = dA + \frac{1}{2}[A \wedge A]$, is the *electromagnetic field*. In local coordinates $F_A = \frac{1}{2}F_{\mu\nu}dx^\mu dx^\nu$. From the *Lagrangian* $\mathcal{L} = -\frac{1}{2}(F_A \wedge *F_A)$ (* the Hodge star operator) we obtain that classical equations of motion $d * F_A = 0$. These together with the *Bianchi identity* $dF_A = 0$ are *Maxwell's equations* (in empty space).

electromagnetism   One of the four *elementary forces* in nature. Classical electrodynamics is governed by *Maxwell's equations* for the *electromagnetic field*.

1-58488-050-3/01/\$0.00+\$.50
© 2003 by CRC Press LLC

element matrix   Given a linear variational problem, based on the sesqui-linear form $a : V \times V \to \mathbb{C}$, and V-conforming *finite element* (K, V_K, Π_K) with *shape functions* $b_1, \cdots, b_M$, $M = \dim V_K$, the corresponding *element matrix* is given by

$$A_K := (a(b_i, b_j))_{i,j=1}^{M}.$$

The bilinear form a could be replaced by a discrete approximation a_h (see *variational crime*).

elementary charge   Electromagnetic fundamental physical constant equal to the charge of a proton and used as *atomic unit* of charge $e = 1.602\ 177\ 33(49) \times 10^{-19}$C.

See *electric charge*.

elementary forces   The four *elementary forces* in nature are *gravitation, electromagnetism, weak nuclear force*, and *strong nuclear force*.

elementary reaction   A reaction in which no reaction intermediates have been detected, or need to be postulated in order to describe the reaction on a molecular scale. Until evidence to the contrary is discovered, an elementary reaction is assumed to occur in a single step and to pass through a single transition state.

elementary symbol   See *semiote*.

elimination   The reverse of an *addition reaction* or *transformation*. In an elimination two groups (called eliminands) are lost most often from two different centers (1/2/elimination or 1/3/elimination, etc.) with concomitant formation of an unsaturation in the molecule (double bond, triple bond) or formation of a new ring.

elliptic equation   A linear *partial differential equation (PDE)* on $\mathbb{R}^n$ of order m with constant coefficients is of the form $\sum_{|j| \le m} a_j D^j u = f$. It is called *elliptic* if the equation in p, $\sum_{|j|=m} a_j p^j = 0$ has no real solution $p \ne 0$.

ellipticity   A *sesqui-linear* form $a : V \times V \to \mathbb{C}$ on a Banach space V is said to be V-elliptic, if

$$|\mathcal{R}a(u, v)| \ge \alpha \|u\|_V^2, \quad \forall u \in V$$

with a constant $\alpha > 0$. An *inf-sup condition* for a is an immediate consequence.

empty collection   For sets, bags, lists, and sequences, the *empty collection* is the corresponding collection which has no elements: thus the empty set, empty list, etc. It is denoted by the corresponding delimiters with nothing between them; thus { }, [ ], and ⟨ ⟩ for empty set, empty list, and empty sequence, respectively.

Comment: If a bag is empty, it reduces to the empty set, also denoted ∅. See also *bag, list, sequence, set*, and *tuple*.

end-group   A *constitutional unit* that is an extremity of a *macromolecule* or *oligomer molecule*.

An end-group is attached to only one *constitutional unit* of a *macromolecule* or *oligomer molecule*.

endomorphism   (**1**) A map from a set to itself, satisfying certain conditions depending on the nature of the set. For example, $f(x * y) = f(x) * f(y)$, if the set is a group.

(**2**) A *morphism* (not necessarily invertible) of an object of a *category* into itself.

energy function   For a *Hamiltonian system*, the *Hamiltonian* is also called the *energy function* of the system.

energy, kinetic   In *classical mechanics*, that part of the energy of a body which the body possesses as a result of its motion. A particle of mass m and speed v has kinetic energy $E = \frac{1}{2}mv^2$.

energy-momentum tensor   In classical field theory invariance of the *Lagrangian* under translations implies, via the *Noether theorem*, conservation of *energy-momentum*. Let the *Lagranian* on space-time be given by $\mathcal{L}(\phi, \partial_\mu \phi)$; then the conserved energy-momentum *tensor* Θ is given by

$$\Theta^{\mu\nu} = \frac{\partial \mathcal{L}}{\partial(\partial_\mu \phi_i)} \partial_\nu \phi_i - g_{\mu\nu} \mathcal{L}, \quad \partial \Theta^{\mu\nu} = 0.$$

In general relativity the *energy-momentum tensor* is defined as the conserved current (via Noether's theorem) given by varying the metric. And in *Yang-Mills theory* (including *Maxwell's equation*) the *energy-momentum tensor* is defined as conserved current obtained by varying the *gauge invariant Lagrangian* with respect to the *Yang-Mills* gauge invariant frame.

entropy   The quantitative measure of disorder, which in turn relates to the thermodynamic functions, temperature, and heat.

enzyme   A catalyst occurring naturally in a biochemical system, or derived from or modeled upon a naturally occurring enzyme. Enzymes are a class of proteins made of polypeptide.

Comment: Until about 15 years ago all known enzymes were proteins, but since then a number of naturally occurring, catalytic RNAs have been discovered. This definition includes the products of laboratory manipulation as well as molecules found in nature.

epimorphism   A *surjective* morphism between objects of a *category*. For example, an *epimorphism* of *vector spaces* is a linear *surjective* map; an *epimorphism* of groups is a *surjective* group homomorphism; an *epimorphism* of *manifolds* is a *surjective* differentiable map. See also *bundle morphism*.

epitope   Any part of a molecule that acts as an antigenic determinant. A *macromolecule* can contain many different *epitopes*, each capable of stimulating production of a different specific *antibody*.

equations of motion   The equations which select the evolution of a mechanical system, or more generally any system with only one independent variable.

See *Hamilton principle* and *Euler-Lagrange equations*.

equicontinuous   Referring to a family F of functions with the property that for every $\epsilon > 0$ there exists a $\delta > 0$ such that $|f(x) - f(y)| < \epsilon$ whenever $|x - y| < \delta$ for all $f \in F$.

equilibrium constant   For any reaction r_i that is at equilibrium and whose solutes are at infinite dilution, $r_i \in R$ (where R is a system of reactions), the *equilibrium constant*, $K_{eq,i}$ is

$$K_{eq,i} = \frac{\prod_{j=1}^{\overline{\mathcal{X}_{d,i}}} x_j^{n_{i,d,j}}}{\prod_{j=1}^{\overline{\mathcal{X}_{s,i}}} x_j^{n_{i,s,j}}} = k_i / k_{-i},$$

where x_j is the concentration of reactant x_j, $n_{i,s|d,j}$ is the stoichiometry of that reactant on

the *sinistralateral* (s) or *dextralateral* (d) side of the reaction equation for reaction r_i, $\overline{\mathcal{X}_{s|d,i}}$ is the number of reactants on the *sinistralateral* and *dextralateral* sides, respectively (| used here as logical "or"), and k_i, k_{-i} are the forward and reverse rate constants, respectively. For the entire system of reactions, $K_{eq,R}$ is

$$K_{eq,R} = \prod_{i=1}^{N} K_{eq,i}.$$

Comment: This definition sets up conditions that allow one reasonably to approximate thermodynamic activities by concentrations, and reactant order by stoichiometries. It also assumes that experiments measuring the apparent *equilibrium constant* have been done by measuring the constant at several different concentrations of reactants and extrapolating the results to zero concentration. Under these conditions, the *equilibrium constant* as used by biochemists comes reasonably close to the *equilibrium constant* as defined in thermodynamics. See also *formal reaction equation*.

equilibrium point   (**1**) A point x_0 is called an *equilibrium point, critical point,* or *singular point* of a *vector field* X if $X(x_0) = 0$. If F_t is the *flow* of X then $F_t(x_0) = x_0$ for all t.

(**2**) (of a *dynamical system* (M, X) A point $x \in M$ such that $X(x) = 0$. The constant curve $c_x(t) = x$ in an *equilibrium point* is an *integral curve*.

equilibrium solution   See *coacervation*.

equivalence classes   See *equivalence relation*.

equivalence relation   A relation $R = \{(a, b) | a, b \in A\}$ that is, $a \sim b \Leftrightarrow (a, b) \in R$ from A to itself such that:

(i.) $\forall a \in A, a \sim a$;

(ii.) $\forall a, b \in A, a \sim b \Rightarrow b \sim a$;

(iii.) $\forall a, b, c \in A, a \sim b, b \sim c \Rightarrow a \sim c$.

If $\sim$ is an *equivalence relation*, the *equivalence class of* $a \in A$ is the subset $[a] = \{b \in A : b \sim a\}$. If $c \in A$ is not in the equivalence class of a (i.e., $c \notin [a]$ or equivalently $a \nsim c$)

then $[c]$ and $[a]$ are disjoint. The set $A/\sim$ of all equivalence classes is a partition of the set A, and it is called the *quotient* of A with respect to the relation $\sim$.

Example: Let $\equiv_n$ be the relation in $\mathbb{Z}$ defined by $a \equiv_n b$ if and only if n divides $a - b$, i.e., if there exists an integer $k \in \mathbb{Z}$ such that $a - b = nk$. It is an equivalence relation. If $a \equiv_n b$ we say that a is congruent to b modulo n. The equivalence classes are $[a]_n = \{a + kn : k \in \mathbb{Z}\}$. The quotient space is denoted by $\mathbb{Z}_n = \{[0]_n, [1]_n, \ldots, [n-1]_n\}$ and it is finite. The structure of additive group of $\mathbb{Z}$ induces a structure of additive group on $\mathbb{Z}_n$ with respect to the operation $[a]_n + [b]_n = [a+b]_n$ which is well defined; i.e., it does not depend on the representatives chosen for the equivalence classes. In fact, if $a' \in [a]_n$ and $b' \in [b]_n$, then $[a']_n + [b']_n = [a' + b']_n = [a + b]_n = [a]_n + [b]_n$.

Specializing to the case $n = 2$, the quotient space $\mathbb{Z}_2$ is made of two elements $[0]_2$ and $[1]_2$. The class $[0]_2$ is the neutral element with respect to the additive structure, while one has $[1]_2 + [1]_2 = [0]_2$. By using the *function* $\exp(i\pi n)$, $\mathbb{Z}_2$ with the additive structure is mapped into the group of signs with the obvious multiplicative group structure.

equivalent norms   Two norms $\|\ \|_1$ and $\|\ \|_2$ on a *normed* vector *space* X such that there is a positive number c such that $c^{-1}\|x\|_1 \leq \|x\|_2 \leq c\|x\|_1$ for all $x \in X$. On *finite* dimensional *vector spaces* all norms are equivalent.

equivariant   See *principal bundle*.

Erlanger program   A plan initiated by Felix Klein in 1872 to describe geometric structures in terms of their groups of *automorphisms*.

essential singularity   A singularity of an *analytic function* that is not a *pole*.

essential spectrum   For a *linear operator* $A : X \to X$ on a normed vector space X the set of scalars λ such that either $\ker(A - \lambda I)$ is infinite dimensional or the image of $A - \lambda I$ fails to be a closed subspace of finite codimension.

essentially self-adjoint operator   See *self-adjoint operator*.

Euler   Leonhard Euler (1707–1783) Swiss mathematician. Some say the most prolific mathematician in history and the first modern mathematician universalist.   He worked at the St. Petersburg Academy and the Berlin Academy of Science.

Euler equations   The motion of a perfect fluid in a domain M (a smooth *Riemannian manifold* with boundary) is governed by the *Euler equations*

$$\begin{cases} \frac{\partial u}{\partial t} + \nabla_u u = -\nabla p \\ div\ u = 0 \end{cases}$$

where u is the velocity field of the fluid and p is the pressure.

The motion of a rigid body with *angular momentum* $\Pi = (\Pi_1, \Pi_2, \Pi_3)$ and principal moments of inertia $\mathbf{I} = (I_1, I_2, I_3)$ is governed by the *Euler equations*

$$\dot{\Pi}_1 = \frac{I_2 - I_3}{I_2 I_3}\Pi_2\Pi_3$$
$$\dot{\Pi}_2 = \frac{I_3 - I_1}{I_1 I_3}\Pi_1\Pi_3$$
$$\dot{\Pi}_3 = \frac{I_1 - I_2}{I_1 I_2}\Pi_1\Pi_2 .$$

Euler-Lagrange equations   The *Lagrangian* formulation of a classical mechanical system described by a *Lagrangian* $L(q^1, \ldots q^n, \dot{q}^1, \ldots \dot{q}^n)$ is given by the principle of least action, which leads to the *Euler-Lagrange equations*

$$\frac{d}{dt}\frac{\partial L}{\partial \dot{q}^i} - \frac{\partial L}{\partial q^i} = 0 , i = 1, \ldots, n.$$

These are equivalent to *Hamilton's equations*.

In field theory where the *Lagrangian* $\mathcal{L}$ depends on fields $\varphi_i(x)$ and their derivatives $\partial_\mu\varphi_i(x)$, the *Euler-Lagrange equations* become

$$\frac{\partial \mathcal{L}(x)}{\partial \varphi_i(x)} - \partial_\mu\frac{\partial \mathcal{L}(x)}{\partial[\partial_\mu\varphi_i(x)]} = 0 , i = 1, \ldots, n.$$

For a mechanical *Lagrangian system* (Q, L) over the configuration space Q with local coordinates q^i the *Lagrangian* is a *function* $L(t, q^i, u^i)$ with (q^i, u^i) local coordinates over the *tangent bundle T Q*. *Euler-Lagrange* equations are given by

$$\begin{cases} u^i = \dot{q}^i \\ \frac{\partial L}{\partial q^i} - \frac{d}{dt}\frac{\partial L}{\partial u^i} = 0 \end{cases}$$

where dot denotes time derivative.

For a *Lagrangian system* $(\mathcal{B}, L)$ of a field theory over the configuration *bundle* $\mathcal{B} = (B, M, \pi, F)$ let us choose local fibered coordinates $(x^\mu, y^i, y^i_\mu, \ldots, y^i_{\mu_1 \ldots \mu_k})$ over the *k-jet* prolongation $J^k\mathcal{B}$ and the *Lagrangian* of the form $L = \mathcal{L}(x^\mu, y^i, y^i_\mu, \ldots, y^i_{\mu_1 \ldots \mu_k})\, ds$, where ds is the standard local volume element of the base *manifold M*. *Euler-Lagrange* equations are given by

$$\begin{cases} y^i_\mu = d_\mu y^i, \ldots, \ y^i_{\mu_1 \ldots \mu_k} = d_{\mu_1 \ldots \mu_k} y^i \\ \frac{\partial L}{\partial y^i} - d_\mu \frac{\partial L}{\partial y^i_\mu} + \cdots + (-1)^k d_{\mu_1 \ldots \mu_k} \\ \frac{\partial L}{\partial y^i_{\mu_1 \ldots \mu_k}} = 0 \end{cases}$$

where d_μ denotes the total derivative with respect to x^μ (i.e., $d_\mu = \partial_\mu + y^i_\mu \partial_i + y^i_{\mu\nu}\partial^\nu_i + \cdots$).

See *Hamilton principle* and *Lagrangian system*.

Euler's formula $\quad e^{i\theta} = \cos\theta + i\sin\theta$ reveals a profound relationship between complex numbers and the trigonometric functions.

Euler's integral $\quad$ The representation for the *gamma function*

$$\Gamma(z) = \int_0^\infty t^{z-1}e^{-t}dt.$$

Euler's method $\quad$ The simplest numerical integration scheme for a differential equation $\dot{x} = f(x)$. The update rule is $x_{n+1} = x_n + f(x_n) + \Delta t$.

evaluation map $\quad$ Let $\mathcal{F}(M)$ denote a set of functions on a *manifold M*. The *evaluation map* is given by $ev : \mathcal{F}(M) \times M \to \mathbb{R} : ev(f, x) = f(x)$, $f \in \mathcal{F}(M), x \in M$.

even function $\quad$ A *function* $f(x)$ such that $f(-x) = f(x)$, for all x. If $g(-x) = -g(x)$, for all x, then g is an *odd function*.

event $\quad$ In probability theory, a measurable set. In relativity, a point in space-time.

evolution equation $\quad$ The *equations of motion* of a dynamical system. They describe the time evolution of a system and are given by differential equations of the form

$$\frac{dx(t)}{dt} = F(x(t)).$$

evolution operator $\quad$ As time passes, the state ψ of a physical system evolves. If the state is ψ_0 at time $t_0 = 0$ and it changes to ψ at a later time t, one sets $F_t(\psi_0) = \psi$. The operator F_t is called the *evolution operator*. Determinism is expressed by the group property $F_t \circ F_s = F_{t+s}$, $F_0 = $ identity.

exact form $\quad$ An *exterior k-form* $\omega \in \Omega^k(M)$ which is the *exterior differential* of a $(k-1)$-form ω, i.e., $\omega = d\theta$. Of course, *exact forms* are *closed*, i.e., $d\omega = 0$. See also *closed form*.

Example: The form $\omega = x dx + y dy$ in $\mathbb{R}^2$ is an exact form, since $\omega = d\left[\frac{1}{2}(x^2 + y^2)\right]$.

excitability $\quad$ The concept of *excitability* first appeared in the literature on neural cells (neurons). It was shown experimentally that a small trigger in the membrane current can lead to large, transient response in membrane electrical potential. This is known as action potential, and it is responsible for the rapid communications between neurons. The mathematical model for this phenomenon was Huxley and it exhibits, among many other features, the threshold phenomenon (see also *threshold phenomenon*).

excluded volume $\quad$ A *polymer* is made of a chain of molecules. These molecules cannot occupy the same position in three-dimensional space. In the simple theory for *polymers* (see *Gaussian chain*), one neglects this effect. A more realistic models for a *polymer* must to consider this effect.

expansion $\quad$ In a network (or graph), the replacement of a smaller subnetwork by a larger one using a sequence of operations on nodes, edges, parameters, and labels.

Comment: In applying it to models of biology such as biochemical networks, expansion requires additional information that specifies which expansion operation, with its components, is appropriate. This is quite distinct from contraction, which needs only its mathematical operations. See *contraction*.

expectation   In probability theory the *expectation* of an random variable f (measurable function) is the integral $\int f \, d\mu$.

exponential   (1) The unique solution to the differential equation $f'(x) = f(x)$, with initial condition $f(0) = 1$. It can also be defined by its power series

$$e^x = 1 + x + \frac{x^2}{2!} + \frac{x^3}{3!} + \cdots$$

(2) The *exponential map* of a *Lie group G* is defined as follows: Let **g** be the *Lie algebra* of G. Any $\xi \in \mathbf{g}$ corresponds to a *left invariant* vector field X_ξ on G. Let c_ξ be the *integral curve* of X_ξ passing through $e \in G, c(0) = e$. The *exponential map* is now defined as exp : $\mathbf{g} \to G : \exp(\xi) = c_\xi(1)$.
Example: For $G = GL(n)$, $\exp(A) = 1 + A + \frac{A^2}{2!} + \frac{A^3}{3!} + \cdots$
(3) The *exponential map* on a *Riemannian manifold* (M, g) is defined as exp : $T_x M \to M :$ $\exp(v) = \gamma(1)$ where γ is the (unique) geodesic with initial conditions (x, v), i.e., $\gamma(0) = x$ and $\dot\gamma(0) = v \in T_x M$.

extension   In this context only, the use of an explicit, fully instantiated, representation of a datum in a database or model.
Comment: The restriction is meant to avoid philosophical wrangling and confine the discourse to that of databases. See also *intension*.

exterior algebra (of a vector space V)   The $\mathbb{Z}$-graded algebra of skew-symmetric covariant tensors on V. If V is finite dimensional, $\{e_i\}$ is a basis of V and $\{e^i\}$ the dual basis in V^* then we denote by $T(V) = \oplus T_k(V)$ the $\mathbb{Z}$-graded algebra of covariant tensors on V. It is infinite dimensional, and it is spanned by

$$\mathbb{I}, \, e^i, \, e^i \otimes e^j, \, \ldots, \, e^{i_1} \otimes \cdots \otimes e^{i_k}, \, \ldots$$

The *exterior algebra* $\Lambda(V) = \oplus \Lambda_k(V)$ is the quotient of this *algebra* by the *bilateral ideal* generated by the elements of the form $e^i \otimes e^j - e^j \otimes e^i$. If V is of dimension $\dim(V) = m$, $\Lambda(V)$ is a finite dimensional $\mathbb{Z}$-graded algebra of dimension 2^m. The induced multiplication $\wedge : \Lambda(V) \times \Lambda(V) \to \Lambda(V)$ is called the *exterior product*. An element in $\Lambda_k(V)$ is called an exterior k-form on V.
Example: Let $\Omega^k(M)$ denote the space of all exterior k-forms on a *manifold M*, $k = 1, \ldots, n = \dim M$. The *exterior algebra* $\Omega(M)$ or *Grassman algebra* of M is the direct sum of the spaces $\Omega^k(M)$, i.e., $\Omega(M) = \oplus_{k=1}^n \Omega^k(M)$.

exterior derivative   A family of operators, on a manifold M, $\mathbf{d}$: $\Omega^k(M) \to \Omega^{k+1}(M)$ ($k = 1, 2, \ldots n = \dim M$), $\Omega^k(M)$ the space of exterior $k - forms$ on M, such that

(i.) $\mathbf{d}$ is a $\wedge$-antiderivative, i.e., $\mathbf{d}(\alpha \wedge \beta) = \mathbf{d}\alpha \wedge \beta + (-1)^k \alpha \wedge \mathbf{d}\beta$, $\alpha \in \Omega^k(M)$;

(ii.) $\mathbf{d}^2 = 0$;

(iii.) for $f \in \Omega^0(M) = C^\infty(M)$ we have $(\mathbf{d}f)_i = \frac{\partial f}{\partial x^i}$.

exterior differential   The derivation d of the exterior algebra of a manifold M defined as follows: if $\omega = \frac{1}{k!}\omega_{\mu_1 \ldots \mu_k}dx^{\mu_1} \wedge \ldots \wedge dx^{\mu_k}$ is a k-form then its exterior differential dω is a $(k + 1)$-form locally expressed as

$$d\omega = \frac{1}{(k+1)!}(k + 1)\partial_{[\mu}\omega_{\mu_1 \ldots \mu_k]}dx^\mu \wedge dx^{\mu_1} \wedge \cdots \wedge dx^{\mu_k}$$

The *exterior algebra* with the exterior differential defines a *cohomology* (i.e., one has d $\circ$ d = 0), and it is called *deRahm* cohomology.
A form ω in the kernel ker(d) (i.e., d$\omega = 0$) is called a *closed form*; a form ω in the image $\Im$(d) (i.e., $\omega = $ dθ for some form θ) is called *exact*. An *exact form* is always closed. The converse is true only provided some topological conditions are met on M. For example, the *Poincaré lemma* proves that all *closed forms* in M are exact when the manifold M is *contractible*.

exterior differential algebra (of a *manifold M*) The $\mathbb{Z}$-*graded algebra* $\Lambda(M) = \oplus_k \Lambda_k(M)$, where $\Lambda_k(M)$ is the set of *k*-forms over M. An element $\omega \in \Lambda_k(M)$ is a skew-symmetric map that associates to *k vector fields* $(X_1, \ldots, X_k)$ a *function* $\omega(X_1, \ldots, X_k)$. Such a map is $\mathcal{F}(M)$-linear in all its arguments, e.g., $\forall f, g \in \mathcal{F}(M)$

$$\omega(f X_1 + g Y_1, \ldots, X_k)$$
$$= f \omega(X_1, \ldots, X_k) + g \omega(Y_1, \ldots, X_k)$$

In other words, a *k*-form ω is a skew-symmetric, $\mathcal{F}(M)$-linear form on the $\mathcal{F}(M)$-module $\mathfrak{X}(M)$ of *vector fields* over M.

exterior form Let $\mathcal{X}(M)$ denote the space of *smooth vector fields* on a *manifold M*. An *exterior k-form* (or exterior differential form of order k) is a *k*-multilinear map $\alpha : \mathcal{X}(M) \times \cdots \times \mathcal{X}(M) \to C^\infty(M)$, k factors, which is skew symmetric, i.e., changes sign if two arguments are interchanged.

exterior product The *exterior product* or *wedge product* of an *exterior k-form* $\alpha \in \Omega^k(M)$ and an *l*-form $\beta \in \Omega^l(M)$ is the $k + l$-form $\alpha \wedge \beta \in \Omega^{k+l}(M)$ defined by

$$\alpha \wedge \beta(X_1, \ldots X_{k+l}) = \sum (sign \; \sigma) \alpha$$
$$\times (X_{\sigma(1)}, \ldots, X_{\sigma(k)}) \beta(X_{\sigma(k+1)}, \ldots X_{\sigma(k+l)}),$$

where the sum is taken over all permutations σ of $\{1, 2, \ldots, k + l\}$.

Basic properties of the $\wedge$-product are

(i.) $\wedge$ is bilinear, i.e., $\alpha \wedge (t\beta + s\gamma) = t(\alpha \wedge \beta) + s(\alpha \wedge \gamma)$, $t, s \in \mathbb{R}$;

(ii.) $\alpha \wedge (\beta \wedge \gamma) = (\alpha \wedge \beta) \wedge \gamma$;

(iii.) $\alpha \wedge \beta = (-1)^{kl} \beta \wedge \alpha$, $\alpha \in \Omega^k(M)$, $\beta \in \Omega^l(M)$.

external photoelectric effect See *photo-emissive detector*.

F

f-functional branch point   See *branch point*.

facilitated diffusion   The Fickian diffusion of a solute, $J = D(c_b - c_a)/(b - a)$, can be enhanced by the presence of a protein which combines reversibly with the solute. The diffusion coefficient of the protein is, of course, significantly lower than that of the solute. Nevertheless, it enhances the overall diffusion of the solute across a distance $b - a$ with concentration difference $c_b - c_a$. The resolute to this counterintuitive phenomenon is that the free solute concentration c_a at a is different from the total solute concentration at a due to the reversible binding. There is an amount of solute bound to the protein. $[PS]_a = Kc_a[P]_a$ where $[P]_a$ is the concentration for unbound protein at a, K is the binding constant. Hence the total solute at a is $(1 + K[P]_a)c_a$ which can be significantly greater than c_a if $K[P]_a \gg 1$. (*cf.* J.D. Murray and J. Wyman, *J. Biol. Chem.*, 246, 5903, 1971).

Faddeev-Popov ghost (not the ghost of Faddeev or Popov)   In quantum field theory with gauge invariant *Lagrangian* one can integrate over the gauge group by fixing a gauge. This leads to the effective *Lagrangian*, which is no longer gauge invariant, by introducing new anticommuting auxiliary scalar fields, so-called *Faddeev-Popov ghosts*. They have similar properties as the Maurer-Cartan form and can be interpreted as *Lie algebra cohomologies*. Integration over such a slice introduces a determinant term change of coordinates which is lifted to the exponent of the action, i.e., added to the *Lagrangian* as a Gaussian integral. This determinant is called the *Faddeev-Popov determinant*. This leads to *BRST* quantization theory. This method of quantization of a system with symmetry is known as the *Faddeev-Popov procedure*.

Fahrenheit scale   A temperature scale devised by Fahrenheit (1686–1736) on which the freezing point of water is 32 degrees and the boiling point of water is 212 degrees, both at standard pressure. (This scale is still used in the U.S.)

faithful action   An *action* $\Phi_g : M \to M$, $g \in G$, of a group G on a space M is called *faithful* or *effective* if $\Phi_9 = id_M$ implies $g = e$.

faradaic current   A current corresponding to the reduction or oxidation of some chemical substance. The net *faradaic current* is the algebraic sum of all the *faradaic currents* flowing through an *indicator* or *working electrode*.

fast Fourier transform (FFT)   A discrete *Fourier transform* algorithm which converts a sampled complex-valued *function* of time into a complex-valued *function* of frequency. It typically reduces the number of computations from order N^2 to $N \log N$.

Fermi-Dirac statistics   In the *Fermi-Dirac statistics* or *quantum statistics*, no more than one of a set of identical particles may occupy a particular quantum state (i.e., the Pauli exclusion principle applies), whereas in the *Bose-Einstein statistics*, the occupation number is not limited. Particles described by these statistics are called *fermions* and *bosons*, respectively.

fermion   A particle described by *Fermi-Dirac statistics*.

Feynman diagram   A sketch in momentum space of the Green's functions describing the interactions of particles. External lines in a diagram represent incoming particles and vertices represent interactions. Straight lines represent electrons and wavy lines photons. Feynman's idea was that *expectation values* of quantities in quantum theory should be obtained by summing the values of these quantities over all histories, i.e., over all diagrams and for each diagram over all positions in space-time for the particle interactions. These summations over all histories are known as *Feynman path integrals*.

Feynman rules   The rules that describe the manipulations with *Feynman diagrams*. These allow physicists to compute Green's functions in quantum field theory.

fiber bundle   A quadruple $\mathcal{B} = (B, M, \pi; F)$ where B, M, and F are differentiable *manifolds* and $\pi : B \to M$ is a *surjective* map of maximal rank. B is called the *total space*, M the *base manifold*, F the *standard fiber*, and π the *projection*. Furthermore, a *trivialization* is required, i.e., a family $\{(U_\alpha, t_\alpha)\}_{\alpha \in I}$ such that:

(i.) U_α are open sets which cover the whole of M, i.e., $\bigcup_{\alpha \in I} U_\alpha = M$;

(ii.) $t_\alpha : \pi^{-1}(U_\alpha) \to U_\alpha \times F$ are diffeomorphisms.

Each (U_α, t_α) is called a *local trivialization*. If a global trivialization $t_\alpha : B \to M \times F$ exists (i.e., if one can take $I = \{1\}$ and $U_1 = \{M\}$) the *bundle* is said to be *trivial*. Trivializations amount to requiring that the total space B of a *bundle* is locally diffeomorphic to the trivial model $M \times F$. Examples: The cylinder $(S^1 \times \mathbb{R}, S^1, p_1, \mathbb{R})$ is a trivial *bundle*, while the *Moebius strip* is fibered over S^1 but it is not trivial, not being globally isomorphic to a Cartesian product.

A preferred class of coordinates $(x^\mu; y^j)$ is then selected on B such that x^μ are coordinates on M and y^j are coordinates on F. These coordinates are called *fibered coordinates*; the *bundle* structure actually amounts to being allowed to separate coordinates in B into two subsets (x^μ and y^j) in a coherent way.

One can restrict trivializations to the fiber $\pi^{-1}(x)$ over a point $x \in U_\alpha$, obtaining diffeomorphisms $t_\alpha(x) : \pi^{-1}(x) \to F$. When two local trivializations are involved (i.e., $x \in U_{\alpha\beta} = U_\alpha \bigcap U_\beta$), we can define the maps $g_{(\alpha\beta)} : F \to F$ given by $g_{(\alpha\beta)} : t_\alpha(x) \circ t_\beta^{-1}(x)$. The maps $g_{(\alpha\beta)} : U_{\alpha\beta} \to \text{Diff}(F)$ are called *transition functions*. They satisfy the *cocycle property*, i.e.,

$$\begin{cases} g_{\alpha\alpha} = \text{id}_F \\ g_{\alpha\beta} = [g_{\beta\alpha}]^{-1} \\ g_{\alpha\beta} \circ g_{\beta\gamma} \circ g_{\gamma\alpha} = \text{id}_F \end{cases}$$

If a family of maps $g_{(\alpha\beta)} : U_{\alpha\beta} \to \text{Diff}(F)$, which satisfies the cocycle property, the base

manifold M and the standard fiber F are provided, then there exists a *bundle* $(B, M, \pi; F)$ (unique up to isomorphisms) (see *bundle morphisms*) having $g_{(\alpha\beta)}$ as transition functions. The total space of this *bundle* is identified with the disjoint union $\coprod_{\alpha \in I} U_\alpha \times F$ [modulo] the equivalence relation

$$(\alpha, x, f) \sim (\beta, y, g)$$
$$\text{if} \quad x = y, g = g_{(\beta\alpha)}(f)$$

Fiber bundles are tightly related to variational calculus. In this framework coordinates x^μ on the base *manifold* are regarded as *independent variables* while y^i are identified with *dependent variables* (i.e., dynamical fields). A configuration is a *(global) section of the bundle*, i.e., a map $\sigma : M \to B$ such that $\pi \circ \sigma = \text{id}_M$.

field   (**1**) In *algebra*, a commutative ring K with at least two elements such that for each $a, b \in K, a \neq 0$ there exists a unique $x \in K$ such that $a \cdot x = b$. Examples are the rational, real, and complex numbers.

(**2**) In physics, functions (real or complex valued) on configuration space are called *scalar fields* and in general, a *field* can be a section of any *vector bundle* over space-time.

field equations   The equations which select the evolution of a field theory.

See *Hamilton principle* and *Euler-Lagrange equations*.

finite element   Let $K \subset \mathbb{R}^n$ be a domain with piecewise *smooth* boundary, and write V_K for a finite dimensional space of piecewise *smooth* functions or *vector fields* on K. If Π_K is a basis of the dual space V'_K, then the triple (K, V_K, Π_K) is a valid finite element. The functionals in the set Π_K are referred to as local degrees of freedom, V_K as local space, and K is called a (geometric) element. For a meaningful finite element the functionals in Π_K have to be *localized* in the sense that they are associated with vertices, edges, faces, etc. of the geometric element. This means that, for $\phi \in \Pi_K$, the value $\phi(v)$, $v \in V_K$, only depends on the restriction of v onto a particular vertex, edge, face, etc.

finite element space   Given a *triangulation* Ω_h of Ω, consider a *finite element* (K, V_K, Π_K) for each cell $K \in \Omega_h$. A related (global) finite element space V_h has to satisfy (for suitable $m \in \mathbb{N}$)

$$V_h \subset V_h^* :=$$

$$\{v : \Omega \to \mathbb{C}^M \text{defined a.e. in } \Omega,$$

$$v_{|K} \in V_K \ \forall K \in \Omega_h\}.$$

The global finite element space is said to be *conforming* with respect to a space V of *functions/vector fields* defined a.e. on Ω, if $V_h \subset V$. Often, this leads to the definition

$$V_h = V_h^* \cap V.$$

In many cases, the requirement of conformity amounts to conditions on the continuity of global finite element *functions* at boundaries between geometric elements. This is due to the fact that *functions* in V^* are piecewise *smooth*. For instance, conformity in $H^1(\Omega)$ entails global continuity, and finite element *functions* in $H^2(\Omega)$ even have to be globally continuously differentiable.

Actually, the glue between the (local) finite elements is provided by the *degrees of freedom*, because in practice V_h is introduced as $V_h := \{v \in V_h^*, \ \forall K, T \in \Omega_h : \phi(v_{|k}) = \kappa(v_{|T})$ if $\phi \in \Pi_K, \kappa \in \Pi_T$ are of the same type and are associated with the same *vertex*, *edge*, *face*, etc. of the global mesh}. It goes without saying that this construction imposes tight constraints on the choice of the local finite elements. Local *degrees of freedom* of the finite elements belonging to adjacent cells have to match.

This makes it possible to convert the local *degrees of freedom* into *nodal degrees of freedom*. These are obtained by collecting all local *degrees of freedom* into one set Π_h and weeding out functionals that agree on V_h. By definition of V_h, the remaining functionals in $\Pi_h := \{\phi_1, \cdots, \phi_N\}$, $N = \dim V_h$, form a basis of the dual space.

finite set   A set whose *cardinality* is either 0 or a natural number. See also *cardinality, countable set, denumerably infinite set, infinite set*, and *uncountably infinite set*.

first integral   For a *dynamical system* (M, X) a *function* $F : M \to \mathbb{R}$ such that $X(F) = 0$. The definition is equivalent to requiring that $F \circ \gamma$ is constant for any *integral curve* γ of X.

More generally, a *function* $F : \mathbb{R} \times M \to \mathbb{R}$ which is constant along the curves of motion. Equivalently, let us define $\hat{X} = \partial_t + X$; F is a *first integral* if $\hat{X}(F) = 0$.

fixed point   A point $x_0 \in M$ is called a *fixed point* of a map $T : M \to M$ if $T(x_0) = x_0$.

flow   The *flow* of a *vector field* X on a *manifold* M is the one-parameter group of *diffeomorphisms* $F_t : M \to M, (F_{t+s} = F_t \circ F_s)$ such that

$$\frac{d}{dt} F_t(x) = X(F_t(x)), \text{ for all } x \in M.$$

In other words, $t \mapsto F_t(x)$ is an *integral curve* (trajectory) of X with initial condition $x \in M$. Flows are also called *dynamical systems*.

flow rate (of a quantity)   Quantity X (e.g., heat, amount, mass, volume) transferred in a time interval divided by that time interval. General symbols: $q_X, \dot{X}$.

fluence, F, Ψ, H_0   At a given point in space, the radiant energy incident on a small sphere divided by the cross-sectional area of that sphere. It is used in photochemistry to specify the energy delivered in a given time interval (for instance, by a laser pulse).

flux   The *flux* of a compound x_j is $\partial x_j / \partial t$ or x_{jt}.

Comment: Biochemically, *flux* is frequently used to describe the consumption of individual molecules by a series of reactions in *dynamic equilibrium*. Experimentally, *flux* can be determined by giving a fixed amount of a particular compound with a radioactive or heavy isotope to a population of cells, then measuring the amount and rate of movement of the isotope into compound(s) derived from the first. Since the cells are usually in a metabolic steady state, the assumption is that all the reactions are in dynamic equilibrium. Compound concentrations do not change over time ($\dot{x}_i = 0$).

Mathematically, the definition of *flux* varies. The alternative taken here is simply to write the ordinary differential velocity equations as partial differential equations, so that $\dot{x}_i$ becomes $\partial x_i/\partial t$ or x_{it}. An alternative is to define *flux* relative to some other compound, which allows one to incorporate the dynamic equilibrium condition directly; the form will vary, much as it does in a physical context. Still another is to use *flux* in other *functions* without defining what it is mathematically. In general, the word is best used only when it is mathematically explicit.

Fock space   In quantum mechanics, the full *Hilbert space* of states.

Focker-Plank equation   Let $U \subset \mathbb{R}^n$ be open and $u : U \times \mathbb{R} \to \mathbb{R}$. The *Focker-Plank equation* for u is

$$u_t - \sum_{i,j=1}^{n}(a^{ij}u)_{x_i x_j} - \sum_{i=1}^{n}(b^i u)_{x_i} = 0.$$

form   See *differential form*.

formal reaction equation   The representation of a chemical or biochemical reaction, the participating molecular species, and the chemical, kinetic, and thermodynamic parameters pertaining to those species in that reaction and to the reaction itself.

The biochemical denotation of formal reaction equations, for example

$$x_1 + x_2 + \ldots \rightleftharpoons x_m + x_{m+1} + \ldots$$

for a set of reactants, $x_j \in \mathcal{X}$, carries with it many less obvious layers of convention, meaning, and information. To make these layers more explicit, the formal reaction equation can be rewritten more systematically as follows. Let the set R be a system of reactions r_i, $1 \leq i \leq N$ among a set of molecular species $\mathcal{X} = \{x_j\}$, $1 \leq j \leq M$ (called *reactants*); i, j, N, and M are any positive integers. Each reactant x_j may participate in more than one reaction, and every reaction has at least two reactants. Each reaction r_i is described by a formal reaction equation of the form

$$\sum_{j=1}^{\overline{\mathcal{X}_{s,i}}} n_{i,s,j} x_j^{k_i} \Leftrightarrow_{k_{-i}} \sum_{j=1}^{\overline{\mathcal{X}_{d,i}}} n_{i,d,j} x_j$$

where the sets of reactants written *sinistralaterally* (s) and *dextralaterally* (d) in the formal reaction equation are $\mathcal{X}_{s,i}$ and $\mathcal{X}_{d,i}$, respectively; $\overline{\mathcal{X}_{s,i}}$ and $\overline{\mathcal{X}_{d,i}}$ are the number of reactants in each set (see *cardinality*); $n_{i,\{s|d\},j}$ is the *stoichiometry* of reactant x_j in $\mathcal{X}_{s,i}$ or $\mathcal{X}_{d,i}$ (| is logical or); and k_i and k_{-i} are the forward and reverse reaction rate constants, respectively. A species appearing catalytically in a reaction equation is included on both sides. R is then described by the set of formal reaction equations, one for each $r_i \in \mathsf{R}$. See also *dextralateral, direction, dynamic equilibrium, microscopic reversibility, product, rate constant, reversibility, sinistralateral, stoichiometry*, and *substrate*.

Fourier   Jean Babtiste Joseph, Baron de Fourier (1768–1830). French analyst and mathematical physicist.

Fourier coefficients   Let $\{x_i\}_{i \in I}$ be an *orthonormal basis* in a *Hilbert space* $(H, <, >)$. Then every $x \in H$ can be written as a Fourier series

$$x = \sum_{i \in I} <x, x_i> x_i.$$

The coefficients $<x, x_i>$ are called the *Fourier coefficients* of x with respect to the basis $\{x_i\}$.

Fourier integral operator   A *Fourier integral operator* A of order k on a compact *manifold* M is locally of the form, for any $u \in C_c^\infty(M)$,

$$Au(x) = (2\pi)^{-n} \int \int e^{i\varphi(x,y,\xi)} a(x, y, \xi)$$
$$\times u(y) dy d\xi,$$

where $a(x, y, \xi)$ is a symbol of order k, and $\varphi(x, y, \xi)$ is a nondegenerate phase function. This defines a bounded linear operator between the Sobolev spaces $A : H_c^s(M) \to H_c^{s-k}(M)$.

Fourier series   See *Fourier coefficients*.

Fourier transform   Let $f \in L^1(\mathbb{R}^n)$. The *Fourier transform* of f, denoted by $\hat{f}$ is the *function* on $\mathbb{R}^n$ defined by

$$\hat{f}(k) = \int_{\mathbb{R}^n} e^{-2\pi i k \cdot x} f(x)\, dx.$$

The *Fourier transform* $f \mapsto \hat{f}$ is norm preserving from $L^2(\mathbb{R}^n)$ to $L^2(\mathbb{R}^n)$.

frame   A set of n linearly independent vectors in the tangent space to an n dimensional *manifold* at a point.

Frechet derivative   Let V, W be *Banach spaces*, $U \subset V$ open. A map $F : U \to W$ is called *Frechet differentiable* at the point $x \in U$ if it can be approximated by a linear map in the form

$$F(x + h) = F(x) + DF(x)h + RF(x, h)$$

where $DF(x)$, called the *Frechet derivative of F* at x, is a bounded (continuous) linear map from V to W, i.e., $DF(x) \in L(V, W)$ and the remainder $RF(x, h)$ satisfies

$$\lim_{h \to 0} \frac{RF(x, h)}{\|h\|} = 0.$$

If the map $DF : U \to L(V, W)$ is continuous in x, then F is called *continuously differentiable* or *of class C^1*. If F is Frechet differentiable at x then F is Gateaux-Levi differentiable at x and $DF(x)h = DF(x, h)$. If F is Gateaux-Levi differentiable at each point $x \in U$ and the map $DF : U \to L(V, W)$ is continuous, then F is Frechet differentiable at each x. In finite dimensions, i.e., $V = \mathbb{R}^n, W = \mathbb{R}^m$, this means $F : \mathbb{R}^n \to \mathbb{R}^m$ is Frechet differentiable if all partial derivatives of F exist and are continuous.

Frechet space   A topolocical *vector space* which is metrizable and complete. Examples are *Banach* and *Hilbert spaces*.

Fredholm alternative   If T is a *compact operator* on a Banach space, then either $(I - T)^{-1}$ exists or $Tx = x$ has a nonzero solution.

Fredholm integral equation   The equation

$$f(x) = u(x) + \int_a^b k(x, t) f(t) dt.$$

Fredholm operator   A bounded linear operator $T : V \to W$ between *Banach spaces V* and W such that

  (i.) $Ker(T) = T^{-1}(0)$ is finite dimensional;

  (ii.) $Rang(T) = T(V)$ is closed;

  (iii.) $W/T(V)$ is finite dimensional.

Example: If $V = W$ and K is a *compact operator*, then $I - K$ is Fredholm.
The integer ind $T = \dim T^{-1}(0) - \dim W/T(V)$ is called the *index of* T.

free action   An *action* $\Phi_g : M \to M, g \in G$, of a group G on a space M which has no fixed points, i.e., $\Phi_g(x) = x$ implies $g = e$.

free Lagrangian   In field theory, a *Lagrangian* without interaction of the involved fields. The corresponding theory is called a *free theory*.

function   An association of exactly one object from a set (the range) with each object from another set (the domain). This is equivalent to defining the *function f* as a set, $f \subseteq A \times B$. For f to be a *function*, it must be the case that if $(x, y) \in f$ and $(x, z) \in f$, then $y = z$.
   Comment: This last condition is equivalent to saying if $a = b$, $f(a) = f(b)$. *Functions* are also called *mappings* and *transformations*. See also *bijection*, *biological functions*, and *relation*.

functional   The term functional is usually used for *functions* whose ranges are $\mathbb{R}$ or $\mathbb{C}$. A *functional f* on a *vector space V* is a *linear functional* if $f(x + y) = f(x) + f(y)$, and $f(\alpha x) = \alpha f(x)$, for all $x, y \in V$ and scalar α.

functional analysis   The theory of infinite dimensional linear algebra, i.e., the theory of topological *vector spaces* (e.g., Hilbert, Banach, or Frechet spaces) and linear operators between them (bounded or unbounded).

functional derivative   Let V be a Banach space, V^* its *dual space*, and $F : V \to \mathbb{R}$ differentiable at $x \in V$. The *functional derivative* of F with respect to x is the unique element $\frac{\delta F}{\delta x} \in V^*$ (if it exists), such that

$$DF(x) \cdot y = < \frac{\delta F}{\delta x}, y >, \text{ for all } y \in V$$

where $<, >: V^* \times V \to \mathbb{R}$ is the pairing between V^* and V and DF is the *Frechet derivative* of F, i.e., $DF(x) \cdot y = \lim_{t \to 0} \frac{1}{t}[F(x + ty) - F(x)]$.

functional motif   A pattern present more than once in a set of biochemical reactions, described from any point of view.

Table 1 Some Examples of *Functional Motives*

Type	An Example
Biochemical	Methyl transfer reaction
Thermodynamic	Reaction r_i has $\Delta G' = G_i$, reaction r_{i+1} has $\Delta G' = G_{i+1}$, $G_i > 0$, $G_{i+1} < 0$, $\|G_{i+1}\| > \|G_i\|$
Chemical	Aldol condensation
Mechanistic	Phosphoenzyme intermediate
Kinetic	Non-allosteric sequential enzyme
Dynamical	Connected reactions exhibiting birythmicity
Topological	Reactions and compounds forming a cycle of length n, $4 \leq n \leq 7$, with at least one reaction requiring an additional compound not a member of the cycle
Regulatory	Rate increased upon binding of ligand
Phylogenetic	Mammalian phosphoglycerate mutases

Comment: Types of *motif* and some examples can be found in Table 1 below. See also *biochemical, chemical, dynamical, kinetic,* mechanistic, phylogenetic, regulatory, *thermodynamic,* and *topological motives.*

functor (**1**) A function between categories.

(**2**) An operator denoting the relation satisfied by a *tuple's* arguments.

Comment: Where the *functor,* also called an operator in some contexts, is written is largely a matter of convention. Some operators are written as prefixes (e.g., derivatives, logical predicates); others are *infix* operators, such as the common arithmetic ones; and still others are *postfix* operators, such as exponentiation. Consider the equation $x = y + a$. This equation uses two binary operators, $=$ and $+$, seen more easily by writing the operations as relations $= (+(y, a), x)$.

fundamental theorem of algebra Every *polynomial* of degree $n \geq 1$ with complex coefficients has at least one root in the complex numbers $\mathbb{C}$.

fundamental theorem of calculus Let f be continuous (hence integrable) on $[a, b]$ and let F be an *antiderivative* of f (i.e., $F'(x) = f(x)$), then

$$\int_a^b f(x)dx = F(b) - F(a).$$

fusion (in biotechnology) The amalgamation of two distinct cells or *macromolecules* into a single integrated unit.

futile cycle A cycle of alternating compound and reactive conjunction nodes which, stoichiometrically, regenerates all compounds in the cycle and consumes more nucleotide or coenzyme molecules than it produces.

Comment: The biochemical connotation of the word is strongly dependent on the notion of futility. A disproportionately large energy or substituent consumption for no apparent synthetic or catabolic change. It also depends on stoichiometry. Clearly for futility to occur, all "nonenergetic" molecules which enter the cycle must remain in it. Thus if some proportion are diverted out of the cycle to other fates, so that the stoichiometry condition is broken, the cycle will decay and energy or substituent consumption will decline.

The quotation marks of "nonenergetic" are meant to warn of elastic biological language. Every molecule has the intrinsic energy of its chemical bonds, so strictly speaking no molecule is nonenergetic. But in a biochemical context, certain bonds of certain molecules such as ATP and NADH are broken in many reactions to yield particularly convenient amounts of energy for the reaction or a substituent group for transfer to another molecule. Compounds used as energy or substituent sources are regenerated by many other reactions. The net result is that energy or substituent groups (or both) are transferred among molecules by these "energetic" or "currency" metabolites.

G

G-graded algebra   An *algebra A* together with a map $\# : g \mapsto A_g$ called the *degree map* which associates a *vector subspace* A_g to any element $g \in G$. The subspaces A_g are not necessarily subalgebras. However, the product of A is *compatible* with the degree map, in the sense that if $a \in A_g$ and $b \in A_h$ then $a \cdot b \in A_{gh}$. The elements $a \in A_g$ are called *homogeneous of degree g*.

Galerkin method   An approach to the discretization of a variational problem

$$u \in V :< A(u), u >= 0 \ \forall v \in V,$$

V a *Banach space*, $A : V \to V'$ continuous. It relies on two finite dimensional subspaces $V_h, W_h \subset V$ to obtain the discrete variational problem

$$u_h \in V_h : \ < A(u_h), v_h >= 0 \ \ \forall v_h \in W_h. \ (4)$$

The space V_h is called the *trial space*, the space W_h is known as the *test space*. If $V_h = W_h$ we confront a *Ritz-Galerkin* method, the more general case is often referred to as a *Petrov-Galerkin method*. Necessary and sufficient conditions for existence and uniqueness of solutions of (4) are supplied *inf-sup conditions*.

Galerkin orthogonality   Let a be a sesquilinear form on a *Banach space* V and V_h be a subspace of V that may represent a *finite element space*. For $f \in V'$ the *functions* $u \in V$ and $u_h \in V_h$ are to satisfy

$$a(u, v) = f(v) \ \ \forall v \in V,$$

$$a(u_h, v_h) = f(v_h) \ \ \forall v_h \in V_h.$$

The *Galerkin orthogonality* refers to the straightforward relationship

$$a(u - u_h, v_h) = 0 \ \forall v_h \in V_h.$$

Galilei group   The closed subgroup of $GL(5, \mathbb{R})$ consisting of Galilei transformations, i.e., of matrices of the following block structure

$$\begin{bmatrix} R & v & a \\ 0 & 1 & t \\ 0 & 0 & 1 \end{bmatrix}$$

where $R \in SO(3)$, $v, a \in \mathbb{R}^3$, $r \in \mathbb{R}$.

gamma function   The complex function given by

$$\Gamma(z) = \int_0^\infty t^{z-1} e^{-t} dt$$

for complex z with positive real part.

Gårding inequality   The inequality satisfied by a *coercive* sesquilinear form on some Sobolev space.

Gateaux derivative   Let V, W be *Banach spaces*, $U \subset V$ open. The *Gateaux derivative* or *directional derivative* of a map $F : U \subset V \to W$ at the point $x \in U$ in direction $h \in V$ is defined by

$$DF(x, h) = \lim_{t \to 0} \frac{1}{t} [F(x + th) - F(x)].$$

F is called *Gateaux differentiable* at $x \in U$ if $DF(x, h)$ exists for all $h \in V$.

Gateaux-Levi derivative   Let V, W be *Banach spaces*, $U \subset V$ open. A map $F : U \to W$ is called *Gateaux-Levi differentiable* at the point $x \in U$ if it is *Gateaux differentiable* at x and the map $h \in V \mapsto DF(x, h)$ from V to W is linear and bounded. If F is *Gateaux-Levi differentiable*, we can set $DF(x)h = DF(x, h)$ and get

$$F(x + h) = F(x) + DF(x)h + RF(x, h)$$

where

$$\lim_{t \to 0} \frac{RF(x, th)}{t} = 0, \ for \ each \ h \in V.$$

gauge group   The group of *gauge transformations*.

gauge invariant   In *field* theory, a *Lagrangian* which admits a *gauge transformation*.

gauge theory   A quantum field theory with a *gauge invariant Lagrangian*.

gauge transformation   In *field* theory a transformation of the fields which leaves the equations of motion invariant. For example, in electrodynamics if we add to the electromagnetic potential A the gradient of a function $\nabla\phi$, then the Maxwell equations are invariant under this *gauge transformation* $A \mapsto A + \nabla\phi$.

In terms of principal fiber bundles a *gauge transformation* is an automorphism of the total space that covers the identity of the base space, i.e.: If $\pi : P \to M$ is a principal G bundle, then a *diffeomorphism* $\phi : P \to P$ is called a *gauge transformation* if $\phi(p \cdot g) = \phi(p) \cdot g$ for all $p \in P, g \in G$ and $\pi(f(p)) = \pi(p)$. *Gauge transformations* form an infinite dimensional *Lie group* (under composition), called the *group of gauge transformations* or *gauge group*. Elements of its *Lie algebra* are called *infinitesimal gauge transformations*.

Gaussian chain   The simplest mathematical model for a *polymer* chain molecule. The model treats the molecule as a random walk and neglects the excluded volume effect. Each step in the walk is assumed to be a random variable. In the limit of a large number of steps, the distribution for the end-to-end distance is approximately Gaussian distributed. (*cf.* P.J. Flory, *Statistical Mechanics of Chain Molecules*, John Wiley & Sons, New York, 1969.)

general linear group   The n^2 dimensional *Lie group* of linear isomorphisms form $\mathbb{R}^n$ to $\mathbb{R}^n$, denoted by $GL(n, \mathbb{R})$. The group of linear isomorphisms form $\mathbb{C}^n$ to $\mathbb{C}^n$ is called the (complex) *general linear group*, denoted by $GL(n, \mathbb{C})$. $A \in GL(n)$ if A is an invertible real (complex) $n \times n$ matrix, i.e., $\det A \neq 0$.

generalized function   Let $\Omega \subset \mathbb{R}^n$ be open and $C_0^\infty(\Omega)$ the *Frechet space* of infinitely differentiable functions with compact support in Ω. A *generalized function* or *distribution* is a continuous linear functional on the space $C_0^\infty(\Omega)$.

geodesic   In *Riemannian geometry* a curve γ whose velocity $\dot\gamma$ is autoparallel to γ, i.e., satisfies the *geodesic equations* $\nabla_{\dot\gamma(t)}\dot\gamma(t) = 0$, or, in local coordinates

$$\ddot\gamma^i + \Gamma^i_{jk}\dot\gamma^j\dot\gamma^k = 0,$$

where Γ^i_{jk} are the *Christoffel symbols*. For example, in $\mathbb{R}^n$ geodesics are straight lines.

More generally, given a *manifold M* with a connection Γ, a curve whose tangent *vector $\dot\gamma$* is parallel transported along the curve itself, i.e., $\nabla_{\dot\gamma}\dot\gamma = 0$.

geodesic equations   See *geodesic*.

geodesic motion   For the kinetic energy *Lagrangian*, with metric tensor g_{ij}

$$L(q^i, \dot q^i) = \frac{1}{2}\sum_{i,j=1}^n g_{ij}(q)\dot q^i \dot q^j$$

the *Euler-Lagrange* equations are called *geodesic flow* or *geodesic motion*; they are equivalent to the *geodesic equations*.

Gibbs energy diagram   A diagram showing the relative standard *Gibbs energies* of reactants, *transition states*, reaction *intermediates* and products, in the same sequence as they occur in a chemical reaction. These points are often connected by a smooth curve (a "Gibbs energy profile," commonly still referred to as a "free energy profile") but experimental observation can provide information on relative standard Gibbs energies only at the maxima and minima and not at the configurations between them. The abscissa expresses the sequence of reactants, products, reaction intermediates, and transition states and is usually undefined or only vaguely defined by the *reaction coordinate* (extent of bond breaking or bond making). In some adaptations the abscissas are, however, explicitly defined as *bond orders*, Bronsted exponents, etc.

Contrary to statements in many textbooks, the highest point on a *Gibbs energy diagram* does not necessarily correspond to the transition state of the *rate-limiting step*. For example, in a *stepwise reaction* consisting of two reaction steps:

(i.) $A + B \rightleftarrows C$;

(ii.) $C + D \to E$.

One of the transition states of the two reaction steps must (in general) have a higher standard Gibbs energy than the other, whatever the concentration of D in the system. However, the value of that concentration will determine which of the reaction steps is rate-limiting. If the particular concentrations of interest, which may vary, are chosen as the standard rate, then the rate-limiting step is the one of highest Gibbs energy.

Gibbs energy of activation (standard free energy of activation), $\Delta^{\ddagger}G^0$ The standard *Gibbs energy* difference between the *transition state* of a reaction (either an *elementary reaction* or a *stepwise reaction*) and the ground state of the reactants. It is calculated from the experimental rate constant k via the conventional form of the absolute rate equation:

$$\Delta^{\ddagger}G = RT[\ln(k_B/h) - \ln(k/T)]$$

where k_B is the Boltzmann constant and h the Planck constant $(k_B/h = 2.08358 \times 10^{10}$ K^{-1}s$^{-1})$. The values of the rate constants, and hence *Gibbs energies of activation*, depend upon the choice of concentration units (or of the thermodynamic standard state).

Gram-Schmidt orthogonalization   A process to construct an *orthonormal basis* in a *Hilbert space* out of an arbitrary Hilbert basis.

Green's functions   Auxiliary functions used to solve nonhomogeneous *boundary value problems*. Example: The general solution of the *boundary value problem*

$$-y'' = f(x) , \ y(0) = 0, \ y(1) = 0$$

can be written in the form

$$y = \phi(x) = \int_0^1 G(x, s) f(s) ds$$

where $G(x, s)$ is the *Green's function* defined by

$$G(x, s) = \begin{cases} s(1 - x), & 0 \le s \le x, \\ x(1 - s), & x \le s \le 1 . \end{cases}$$

Green's theorem   A special case of Stokes' theorem for the plane. Let P, Q be differentiable functions in a region $\Omega \subset \mathbb{R}^2$, then

$$\int \int_\Omega \left(\frac{\partial Q}{\partial x} - \frac{\partial P}{\partial y} \right) = \int_{\partial \Omega} P dx + Q dy.$$

grammar   For any language $\mathcal{L}$, the *grammar* Π is the set of rules specifying the syntax of well-formed constructs in $\mathcal{L}$.

Comment: A synonym for the rules (and confusingly, sentences formed by applying the rules) is "productions." Grammars have three functions: to generate and to recognize constructs in a language, and to transform one language to another. The most familar example of a grammar comes from string grammars built from natural languages, which specify the syntactic properties a sentence must fulfill for it to be "legal." Many grammars, and the languages they describe, fall into a hierarchy of increasing mathematical complexity first devised by Noam Chomsky. A context-sensitive grammar, one of the more complex types, specifies that a token's output depends on its context. Examples in English are a little contrived: perhaps the best is "Dick and Jane went north and south, respectively." Here "respectively" signals a mapping function, so that Dick went north and Jane south. Grammars are commonly applied to recognize features of DNA and protein sequence. In that context they are usually called string grammars. They are also used to recognize and generate patterns of chemical and biochemical structure and function. See *graph grammar*.

graph   A *graph* $\mathsf{G}(\mathcal{V}, \mathcal{E})$ consists of a set of *vertices* $\mathcal{V}$, $\mathcal{V} \neq \emptyset$ and a set of *edges* $e(\lambda, v_i, v_j) \in \mathcal{E}, \mathcal{E} \ge \emptyset$, where $\lambda \in \Lambda, \Lambda \neq \emptyset$ is the type of relation the *edge* expresses, and $\{v_i, v_j\} \in \mathcal{V}, i \neq j$ are the (possibly empty) vertices associated with that *edge*.

Comment: This definition has *edges* expressing relationships but allows them to be unbounded by vertices on either or both sides ("free" edges). This latter feature is particularly useful in specifying certain types of graphs and operators upon them. All the graphs considered here are finite; have one and only one edge joining any pair of nodes (are not *multigraphs*); and have no edges with only one node (*loops*; not a *pseudograph*). Derived from G. Rozenberg (personal communication).

graph grammar   A *grammar* over a set $\mathcal{G}$ of graphs $G(\mathcal{V}, \mathcal{E})$.

Comment: Like a natural language or string grammar, a *graph grammar* provides rules to generate, recognize, and transform graphs; for example, expanding a token for a substituent group into the atoms and bonds of that group, or contracting the group to an abbreviation. Mirroring the chemistry, one could form bonds and specify orientation by examining an atom's context. This suggests the *grammar* is context-sensitive. However, to date the only formal proofs for molecular languages are that linear *polymers* such as DNA are greater than context-free (noncontext-free), a superset which includes the context-sensitive grammars among its members. So while it is reasonable to conjecture that molecules form a context-sensitive language, there is no formal proof yet.

graph theory   The branch of mathematics dealing with the topological relationships among abstract graphs.

Comment: Graph theory is a rich source for pattern recognition algorithms.

Grassman algebra   See *exterior algebra*.

ground   In logic and logic programming, a term which is not, or does not include, a variable.

Comment: The restriction is meant to avoid wiring confusion.

group   (**1**) A defined linked collection of atoms or a single atom within a *molecular entity*. The use of the term in physical organic and general chemistry is less restrictive than the definition adopted for the purpose of nomenclature of organic compounds.

(**2**) A set $\mathcal{G}$ with a binary operation which is associative. Each element is assumed to have an inverse and $\mathcal{G}$ contains an *identity* element.

group homomorphism   A map $\phi : G \rightarrow H$ between two groups such that

$$\phi(e_G) = e_H \qquad \phi(g_1 \cdot g_2) = \phi(g_1) \cdot \phi(g_2)$$

where e_G and e_H denote the identity elements of the groups G and H, respectively.

H

h-version of finite elements   A strategy that seeks to achieve a sufficiently small discretization error of a finite element scheme for a boundary value problem by using fine meshes. This can mean a small global *meshwidth* or local refinement. The latter is employed in the context of *adaptive refinement*.

half-life   The time $t_{1/2}$ required for one half of a population to change randomly from some state $\int$ to another state $\int'$. At least one of these two states must be observable.

Comment: The classic example of half-life is radioactivity. In any sample of matter, some of the nuclei will be radioactive isotopes and the remainder not. The decline in the radioactivity of the sample is governed by a first-order Poisson process

$$N = N_0 e^{-\lambda t}$$

where[1] N and N_0 are the current and original number of radioactive atoms in the sample and λ the characteristic decay constant for an isotope. Thus $t_{1/2} = \ln 2/\lambda$. Judged by the metric of radioactivity, nonradioactivity is a nonobservable state (though it can of course be observed by other assays).

Hamilton equations   Let $(q^1, \ldots, q^n, p_1, \ldots, p_n)$ be *canonical coordinates* and H a *smooth function*. *Hamilton's equations* for H are

$$\dot{q}^i = \frac{\partial H}{\partial \dot{p}_i}, \quad \dot{p}_i = -\frac{\partial H}{\partial \dot{q}^i}, \quad i = 1, \ldots, n.$$

Hamilton-Jacobi equation   Let $U \subset \mathbb{R}^n$ be open and $u : U \times \mathbb{R} \to \mathbb{R}$. The *Hamilton-Jacobi equation* for u is

$$u_t + H(Du, x) = 0,$$

where $Du = D_x u = (u_{x_1}, \ldots, u_{x_n})$ denotes the gradient of u with respect to the spatial variable $x = (x_1, \ldots, x_n)$.

---

[1]This notation is restricted to this discussion and characteristic of that seen in other texts on this subject; do not confuse it with notation on reactions elsewhere, especially the use of N and λ.

Hamilton principle   The principle according to which the *action functional* determines *configurations of motion* (also called *critical configurations*). Critical configurations are those for which the action functional is stationary with respect to all compactly supported deformations. For this reason, the *Hamilton principle* is also called the *principle of stationary action*.

In *Lagrangian systems* the Hamilton principle is equivalent to the *Euler-Lagrange equations*, which, depending on the number of independent variables ($n = 1$ or $n > 1$), are called *equations of motion* or *field equations*, respectively.

See also *Lagrangian system* and *action functional*.

Hamiltonian system   Let (M, ω) be a *symplectic manifold* and $H : M \to \mathbb{R}$ a *smooth function*. The corresponding *Hamiltonian vector field* X_H is determined by the condition $\omega(X_H, Y) = dH \cdot Y$. The *flow of X_H* in *canonical coordinates* satisfies *Hamilton's equations* and is called a *Hamiltonian system*.

More generally, an *evolution equation* is called a *Hamiltonian system* if it can be written in the form $\dot{F} = \{F, H\}$ with respect to some *Poisson bracket* $\{\,,\,\}$. H is called the *Hamiltonian* of the system.

Hamiltonian vector field   Let $(P, \{\,,\,\})$ be *Poisson manifold* and $H \in C^\infty(P)$. The *Hamiltonian vector field* X_H of H is defined by

$$X_H(G) = \{H, G\}, \text{ for all } G \in C^\infty(P).$$

Hammett equation (Hammett relation)   The equation in the form:

$$\lg(k/k_0) = \rho\sigma$$

or

$$\lg(K/K_0) = \rho\sigma$$

applied to the influence of *meta-* or *para-* substituents X on the reactivity of the functional group Y in the benzene derivative m- or p-XC_6H_4Y. k or K is the rate or equilibrium constant, respectively, for the given reaction of m- or p-XC_6H^4Y; k_0 or K_0 refers to the reaction of C_6H_5Y, i.e., $X = H$; σ is the substituent constant characteristic of m- or p-X: ρ is the reaction constant characteristic of the given reaction of Y. The equation is often encountered in a form with $\lg k_0$ or $\lg K_0$ written as a separate term on the right-hand side, e.g.,

$$\lg k = \rho\sigma + \lg k_0$$

or

$$\lg K = \rho\sigma + \lg K_0$$

It then signifies the intercept corresponding to $X = H$ in a regression of $\lg k$ or $\lg K$ on σ.

See also *Yukawa-Tsuno equation*.

harmonic form   A k-form α on a *manifold* M is called *harmonic* if $\Delta\alpha = 0$, where $\Delta = d\delta + \delta d$ is the Laplace-deRham operator on M.

harmonic frequency generation   Production of coherent radiation of frequency $k\nu(k = 2, 3, \ldots)$ from coherent radiation of frequency ν. In general, this effect is obtained through the interaction of laser light with a suitable optical medium with nonlinear polarizability. The case $k = 2$ is referred to as frequency doubling, $k = 3$ is frequency tripling, and $k = 4$ is frequency quadrupling. Even higher integer values of k are possible.

harmonic function   A function u satisfying $\Delta u = 0$. See *Laplace equation*.

harmonic oscillator   The *function* $H = \frac{1}{2}(-d^2/dx^2 + x^2)$ is the *Hamiltonian* of the *harmonic oscillator* (assuming $m = 1$). The integral curves (trajectories) are circles.

heat, q, Q   Energy transferred from a hotter to a cooler body due to a temperature gradient.

heat equation   Let $U \subset \mathbb{R}^n$ open and $u : U \times \mathbb{R} \to \mathbb{R}$. The *heat equation* or *diffusion equation* for u is

$$u_t - \Delta u = 0.$$

heat kernel   On $\mathbb{R}^n \times \mathbb{R}^n$ the function

$$e^{t\Delta}(x, y) = (4\pi t)^{-n/2}\exp\left\{-\frac{|x - y|^2}{4t}\right\}$$

The action of the *heat kernel* on a *function* f is defined by

$$(e^{t\Delta}f)(x) = \int_{\mathbb{R}^n} e^{t\Delta}(x, y)f(y)dy.$$

Heisenberg uncertainty principle   In quantum mechanics, a principle set forth by Heisenberg which asserts that the simultaneous exact measurements of the values of position and momentum is impossible. If Δq is the range of values found for the coordinate q of a particle, and Δp is the range in the simultaneous measurement of the corresponding momentum, then $\Delta q \cdot \Delta p \geq h$, where h is the Planck constant.

helicity   A quantitative measure of the amount of helix in a *polymer* molecule with helical structure.

helix   In biochemistry, a molecular structure having a given skew symmetry. For proteins, there is α-helix which was first proposed by L. Pauling. For DNAs, there is a double helix.

helix-coil transition   A mathematical model originally develped for the probability distribution of α-helix formation in a polypeptide. The model is formulated based on the transfer matrix method. It is a generalization of *one-dimensional Ising model* (*cf.* D. Poland and H.A. Scheraga, *Theory of Helix-Coil Transitions*, Academic Press, New York, 1970).

Helmholtz equation   Let $U \subset \mathbb{R}^n$ be open and $u : U \subset \mathbb{R}^n \to \mathbb{R}$. The *Helmholtz equation* or *eigenvalue equation* for u is

$$-\Delta u = \lambda u.$$

Hermitian operator   See *self-adjoint operator*.

heterolysis (heterolytic)   The cleavage of a *covalent bond* so that both bonding electrons remain with one of the two fragments between which the bond is broken.

Higgs mechanism   In quantum field theory, the spontaneous breaking of gauge symmetry, in which a massless gauge boson (Goldstone boson) and a massless scalar field combine to form a massive gauge boson, called a *Higgs boson*.

Hilbert-Schmidt operator   A *bounded linear operator* T on a *Hilbert space* H is called *Hilbert-Schmidt* if $\mathrm{trace}\,T^*T < \infty$, that is, $\sum \lambda_j < \infty$ for the eigenvalues of T^*T.

Hilbert space   A *vector space H* which has an *inner product* $< \ , \ >$ (scalar product) and is *complete* with respect to the induced norm $\|x\|^2 = < x, x >$.

Hodgkin-Huxley model   A mathematical model for the dynamics of electrical potential and ionic currents, due to sodium and potassium, across a biological cell membrane. It consists of four nonlinear ordinary differential equations. The model exhibits various interesting behavioral characteristics observed experimentally such as threshold phenomenon and oscillation. See *threshold phenomenon* and *excitability*.

Hölder inequality   Let p, q, r be positive integers satisfying $p, q, r \geq 1$ and $p^{-1} + q^{-1} = r^{-1}$. If $f \in L^p(X, d\mu)$, $g \in L^q(X, d\mu)$, then $fg \in L^r(X, d\mu)$ and *Hölder's inequality* holds:

$$\|fg\|_r \leq \|f\|_p \|g\|_q$$

homeomorphism (between two topological spaces X and Y)   A map $\varphi : X \to Y$ which is continuous with a continuous inverse map $\varphi^{-1} : Y \to X$.

homolysis (homolytic)   The cleavage of a bond ("homolytic cleavage" or "homolytic fission") so that each of the molecular fragments between which the bond is broken retains one of the bonding electrons. A *unimolecular* reaction involving *homolysis* of a bond (not forming part of a cyclic structure) in a molecular entity containing an even number of (paired) electrons results in the formation of two radicals.

It is the reverse of *colligation*. *Homolysis* is also commonly a feature of *bimolecular substitution reactions* (and of other reactions) involving radicals and molecules.

homotopy   Let X and Y be *topological spaces* and $\varphi : X \to Y$ and $\phi : X \to Y$ be two continuous maps from X to Y. They are *homotopic* if there exists a continuous map $F : [0, 1] \times X \to Y$ such that

(i.) $F(0, x) = \varphi(x)$;

(ii.) $F(1, x) = \phi(x)$.

The map F is called a *homotopy* between φ and ϕ.

Let $A \subset X$ be a subset; a *homotopy F* between the maps φ and ϕ is a *homotopy relative to $A \subset X$* if we have also:

(iii.) $\forall t \in [0, 1]$, $\forall a \in A$, $F(t, a) = \phi(a) = \varphi(a)$.

In particular one can consider *homotopy of loops based at $x_0 \in X$* in a topological space X. *Homotopy* relative to $A = \{x_0\}$ is an *equivalence relation* on the set of all loops based at $x_0 \in X$ and the quotient space $\pi_0(X, x_0)$ is called the *homotopy group of X*. It is in fact a group under the compositions induced by loop composition (see *loop*); this group does not depend on the base point $x_0 \in X$, and it is a *topological invariant* of X.

See also *contractible*.

horizontal lift (induced by a connection Δ_b) The unique *vector* $\Gamma(X) \in \Delta_b$ projecting onto X. Local generators of the space Δ_b are of the form $\partial_\mu - \Gamma_\mu^i(x, y)\partial_i$. If $X = X^\mu \partial_\mu$ is a vector field over M, then the horizontal lift of X is locally given by

$$\Gamma(X) = X^\mu (\partial_\mu - \Gamma_\mu^i(x, y))\partial_i \in \Delta_b$$

hydrocarbons   Compounds consisting of carbon and hydrogen only.

hydron   The general name for the cation H^+; the species H^- is the hydride anion and H is the hydro group. These are general names to be used without regard to the nuclear mass of the hydrogen entity, either for hydrogen in its natural abundance or where it is not desired to distinguish between the isotopes.

hyperbola   The conic section with equation $x^2/a^2 - y^2/b^2 = 1$. See *asymptote to the hyperbola*.

hyperbolic critical point   A *critical point* m_0 of a vector field X (i.e., $X(m_0) = 0$) is called *hyperbolic* or *elementary* if none of the *eigenvalues* of the linearization $X'(m_0)$ (called *characteristic exponents*) has zero real part. Liapunov's theorem shows that near a hyperbolic critical point the flow looks like that of its linearization.

hyperbolic equation   A second-order *partial differential equation* of the form

$$Au_{xx} + Bu_{xy} + Cu_{yy} + Du_x + Eu_y + Fu = G$$

such that $B^2 - 4AC > 0$. It is called *parabolic* if $B^2 - 4AC = 0$ and *elliptic* $B^2 - 4AC < 0$.

I

ideal (of an algebra A)   A *left ideal* is a *vector subspace* $I \subset A$ such that $\forall a \in A,\ i \in I,\ a \cdot i \in I$. Analogously, a *right ideal* is a *vector subspace* $I \subset A$ such that $\forall a \in A,\ i \in I,\ i \cdot a \in I$. A *bilateral ideal* is left ideal which is also a right ideal.

The quotient of an *algebra* by a bilateral ideal is an *algebra*. In the category of algebras with unity, proper ideals are not subalgebras. In fact, if the unit $1 \in I$ belongs to the *ideal*, then $I = A$. On the contrary, *ideals* of *Lie algebras* are always subalgebras.

identity   (**1**) An element e of a set X with a binary operation * satisfying

$$a * e = e * a = a$$

for all $a \in X$.

(**2**) A true mathematical equation.

image   Let r be a relation with domain A and codomain B, and let $a \in A,\ r(a) \in B$. Then the *image* of a under r, $r(a) \in B$, is produced by applying r to a.

Comment: See the comment on *relation* for more details. See also *codomain, domain, range*, and *relation*.

immersional wetting   A process in which a solid or liquid, β, is covered with a liquid, α, both of which were initially in contact with a gas or liquid, δ, without changing the area of the $\alpha\delta$-interface.

immunoglobulin (Ig)   A protein of the globulin-type found in serum or other body fluids that possesses *antibody* activity. An individual Ig molecule is built up from two light (L) and two heavy (H) polypeptide chains linked together by disulfide bonds. Igs are divided into five classes based on antigenic and structural differences in the H chains.

implicit function theorem   Let V, W, Z be *Banach spaces*, $U_1 \subset V, U_2 \subset W$ open and $f : U_1 \times U_2 \to Z$ differentiable. For some $x_0 \in U_1,\ y_0 \in U_2$ assume $D_2 f(x_0, y_0) : W \to Z$ is an *isomorphism*. Then there are *neighborhoods* U_0 of x_0 and Z_0 of $f(x_0, y_0)$ and a unique differentiable map $g : U_0 \times Z_0 \to U_2$ such that for all $(x, z) \in U_0 \times Z_0$

$$f(x, g(x, z)) = z.$$

incidence relation   See *edge*.

incident   A node v_i is *incident* to an *edge* (v_i, v_j), since it is an end point of the *edge*.

index   See *Fredholm operator, Atiyah-Singer index theorem*.

inf-sup condition   A *sesqui-linear* form $a : V \times V \to \mathbb{C}$ on a *Hilbert space* V satisfies an inf-sup condition if, for some $\alpha > 0$,

$$\sup_{v \in V} \frac{|a(u, v)|}{\|v\|_V} \geq \alpha \|u\|_V \quad \forall u \in V,$$

$$\sup_{v \in V} \frac{|a(u, v)|}{\|v\|_V} > 0 \quad \forall u \in V.$$

This is necessary and sufficient for the variational problem

$$u \in V : a(u, v) = f(v) \quad \forall v \in V$$

to possess a unique solution for any $f \in V'$. The solution satisfies $\|u\|_V \leq \alpha^{-1} \|f\|_{V'}$.

For a linear symmetric variational *saddle point problem* with sesqui-linear forms $a : V \times V \to \mathbb{C}$ and $b : V \times W \to \mathbb{C}$ the suitable inf-sup conditions claim the existence of constants $\alpha, \beta > 0$ such that

$$\mathcal{R}\{a(u, u)\} \geq \alpha \|u\|_V^2 \quad \forall u \in \text{Ker}(B),$$

$$\sup_{v \in V} \frac{|b(v, p)|}{\|v\|_V} \geq \beta \|p\|_V \quad \forall p \in V$$

where

$$\text{Ker}(B) := \{v \in V : b(v, q) = 0 \ \forall q \in W\}.$$

Then the variational saddle point problem, which seeks $u \in V,\ p \in W$ such that

$$a(u, v) + b(v, p) = f(v) \quad \forall v \in V$$

$$b(u, q) = g(q) \quad \forall q \in W,$$

has a unique solution. It satisfies

$$\|u\|_V \leq \frac{1}{\alpha} \|f\|_{V'} + \left(\frac{\|a\|}{\alpha} + 1 \right) \frac{1}{\beta} \|g\|_{W'},$$

$$\|p\|_W \leq$$
$$\frac{1}{\beta} \left(\frac{\|a\|}{\alpha} + 1 \right) \|f\|_{V'} + \left(\frac{\|a\|}{\alpha} + 1 \right) \frac{\|a\|}{\beta^2} \|g\|_{W'}.$$

Similar results are known for nonsymmetric saddle point problems.

infinite set   A set whose *cardinality* is not finite. See also *cardinality, countable set, denumerably infinite set, finite set*, and *uncountably infinite set*.

infix   An operator written between its operands; thus for two operands x and y and *operator* f, the syntax is xfy. See also *functor, postfix, prefix*, and *relation*.

inherent viscosity (of a polymer)   The ratio of the natural logarithm of the *relative viscosity*, η_r, to the mass concentration of the polymer, c, i.e.,

$$\eta_{inh} \equiv \eta_{\ln} = (\ln \eta_r)/c.$$

The quantity $\eta_{\ln}$, with which the inherent viscosity is synonymous, is the logarithmic viscosity number.
Notes: (**1**) The unit must be specified: $cm^3 g^{-1}$ is recommended.
(**2**) These quantities are neither viscosities nor pure numbers. The terms are to be looked at as traditional names. Any replacement by consistent terminology would produce unnecessary confusion in the *polymer* literature.

inhibitor   A substance that diminishes the rate of a chemical reaction; the process is called *inhibition*. Inhibitors are sometimes called negative catalysts, but since the action of an inhibitor is fundamentally different from that of a *catalyst*, this terminology is discouraged. In contrast to a catalyst, an inhibitor may be consumed during the course of a reaction. In enzyme-catalyzed reactions an *inhibitor* frequently acts by binding to the enzyme, in which case it may be called an *enzyme inhibitor*.

injection   A map $\phi : A \to B$ which is *injective*, so that whenever $\phi(a_1) = \phi(a_2)$ then $a_1 = a_2$.

injective   Let A and B be two sets, with A the domain and B the codomain of a *function* f. Then the *function* f is *injective* if, for any x and $x' \in A, x \neq x'$, their images $f(x) \neq f(x')$. Injective functions are also called *one-to-one transformations* of A into B. See also *into, onto, bijection*, and *surjection*.

inner product   An *inner product* or *scalar product* on a *complex vector space* V is a map $< \, , \, >: V \times V \to \mathbb{C}$ such that, for all $x, y, z \in V, \alpha \in \mathbb{C}$

(i.) $< x, y > = \overline{< y, x >}$ (the bar denotes complex conjugate);

(ii.) $< x + y, z > = < x, z > + < y, z >$;

(iii.) $< \alpha x, y > = \alpha < x, y >$;

(iv.) $< x, x > \geq 0$;

(v.) $< x, x > = 0$ only if $x = 0$.

Such a space is called an *inner product space*.

inner product operator   Let M be a *manifold*, $\Omega^k(M)$ the space of *exterior k-forms* on M and X a vector field on M. The *inner product operator* $\mathbf{i}_X$ is the linear map $\mathbf{i}_X : \Omega^{k+1}(M) \to \Omega^k(M)$ defined by

$$\mathbf{i}_X \omega(X_1, \ldots, X_k) = \omega(X, X_1, \ldots, X_k),$$

$\omega \in \Omega^{k+1}(M)$, $X_1, \ldots X_k$ vector fields. $\mathbf{i}_X$ is an antiderivative with respect to the $\wedge$ product.

instance   In logic and logic programming, A is an *instance* of B if there exists a substitution θ such that $A = B\theta$. See also *instantiation*.

instantiation   In logic and logic programming, the substitution of a ground term for a variable to produce an instance of the variable.
Comment: Substitution has specialized meanings in chemistry and logic programming; here the logic programming meaning is used. See also *instance*.

integrable distribution   A *distribution of subspaces* Δ such that for each point $x \in M$ there exists an *integral manifold* through x.

integral (sub)manifold (of a distribution of subspaces Δ of rank n)   A submanifold $N \subset M$ such that $T_x N \subset \Delta_x$ for all $x \in N$. Usually N is required to be of dimension $n = \text{rank}(\Delta)$

integrable system   A *Hamiltonian system* X_H on a $2n$-dimensional *symplectic manifold* $(M\omega)$ such that there exist n linearly independent functions $H = F_1, \ldots, F_n$ in *involution*, i.e., $\{F_i, F_j\} = 0$, $i, j = 1, \ldots, n$ and $dF_i(m)$ are linearly independent at each point $m \in M$.

integral curve   Let X be a *vector field* on a *manifold M*. An *integral curve* of X with initial condition x_0 at $t = 0$, is a *smooth curve* $c : [a, b] \to M$ such that $c(0) = x_0$ and

$$c'(t) = X(c(t)), \quad \text{for all } t \in [a, b].$$

See also *dynamical system*.

intension   In this context only, the use of an implicit representation of a datum in a database or model, which is not fully instantiated within the database and whose value must be computed according to the representation.

Comment: The restriction is meant to avoid philosophical wrangling and confine the discourse to that of databases. A common intensional representation is a declarative rule, but the output of a statistical algorithm, federated to a database and performed on data derived from it or other computations, is another example. See also *extension*.

interior (of a set)   For a subset A of a topological space, the set of points $p \in A$ such that A also contains a *neighborhood* of p.

intermediate   A *molecular entity* with a *lifetime* appreciably longer than a molecular vibration (corresponding to a local potential energy minimum of depth greater than RT) that is formed (directly or indirectly) from the reactants and reacts further to give (either directly or indirectly) the products of a *chemical reaction*; also the corresponding *chemical species*.

internal energy, U   Quantity of change which is equal to the sum of heat, q, brought to the system and work, w, performed on it, $\Delta U = q + w$. Also called thermodynamic energy.

into   In mathematics, sometimes the word is used to identify a one-to-one mapping, sometimes simply a mapping of a set X which transforms points of X into points of another set Y, e.g., $y = x^2$ is a mapping of the real numbers into the real numbers, or onto the nonnegative real numbers. See also *onto, bijection, injection,* and *surjection*.

intramolecular   (**1**) Descriptive of any process that involves a transfer (of atoms, groups, electrons, etc.) or interactions between different parts of the same *molecular entity*.

(**2**) Relating to a comparison between atoms or groups within the same molecular entity.

intramolecular isotope effect   A *kinetic isotope effect* observed when a single substrate, in which the isotopic atoms occupy equivalent reactive positions, reacts to produce a nonstatistical distribution of *isotopomeric* products. In such a case the *isotope effect* will favor the pathway with lower force constants for displacement of the isotopic nuclei in the *transition state*.

intrinsic rate constant   See *rate constant*.

intrinsic viscosity (of a polymer)   The limiting value of the *reduced viscosity*, η_i/c, or the *inherent viscosity*, η_{inh}, at infiinite dilution of the polymer, i.e.,

$$[\eta] = \lim_{c \to 0} (\eta_i/c) = \lim_{c \to 0} \eta_{inh}$$

Notes: (**1**) This term is also known in the literature as the Staudinger index.

(**2**) The unit must be specified; $cm^3 g^{-1}$ is recommended.

(**3**) This quantity is neither a viscosity nor a pure number. The term is to be looked on as a traditional name. Any replacement by consistent terminology would produce unnecessary confusion in the *polymer* literature. Synonymous with limiting viscosity number.

inverse function theorem   Let V, W be *Banach spaces*, $U \subset V$ open and $f : U \subset V \to W$ differentiable. If $Df(x_0)$ is a linear isomorphism $Df(x_0) : V \to W$, then f is a *local diffeomorphism*, i.e., a *diffeomorphism* from a *neighborhood* of x_0 onto a *neighborhood* of $f(x_0)$.

inverse inequality   Inequalities bounding stronger norms (in the sense that they contain higher derivatives) of *functions* in finite dimensional spaces by weaker norms. An important example are bounds for the $H^1(\Omega)$-norm of finite element functions in terms of their $L^2(\Omega)$-norm. The constants in these inequalities are invariably for families of approximating spaces that are *asymptotically dense*. For instance, for the above pair of norms and *h-version* families of $H^1(\Omega)$-conforming *finite element spaces* on *quasi-uniform* and *shape regular* meshes the constant will grow like $O(h^{-1})$ where h is the *mesh width*.

inverse relation   If $R \subseteq A \times B$ is a relation from A to B, then the set $R^{-1} = \{(b, a) \mid (a, b) \in R\}$ is a subset of $B \times A$. R^{-1} is called the *inverse* of R.

invertible   A map $\phi : A \to B$ is *invertible* if there exists a map $\phi^{-1} : B \to A$ such that

$$\phi^{-1} \circ \phi = \mathrm{id}_A \qquad \phi \circ \phi^{-1} = \mathrm{id}_B$$

The map ϕ^{-1} is called the *inverse* of ϕ.

involution of functions   Two functions f and g on a symplectic *manifold* are in *involution* when their *Poisson bracket* vanishes, i.e., $\{f, g\} = 0$.

involutive distribution   A distribution of subspaces Δ on M with generators X_A such that $[X_A, X_B] \in \Delta$ where $[,]$ is the *commutator*. A distribution is *involutive* if and only if it is integrable (Frobenius theorem).

ion channel   In biochemistry, a protein, embedded in cell membrane, which conducts ionic current of specific ions. An *ion channel* can be either passive or modulated by the electrical voltage across the cell membrane. The latter is often called voltage-gated ion channel. (*cf.* B. Hille, *Ionic Channels of Excitable Membranes*, 2nd ed., Sinauer Associates, Sunderland, MA, 1992).

ion pair   A pair of oppositely charged ions held together by Coulomb attraction without formation of a *covalent bond*. Experimentally, an *ion pair* behaves as one unit in determining conductivity, kinetic behavior, osmotic properties, etc.

Following Bjerrum, oppositely charged ions with their centers closer together than a distance

$$q = 8.36 \times 10^6 z^+ z^- /(\epsilon_r T)\mathrm{pm}$$

are considered to constitute an *ion pair* ("Bjerrum ion pair"). [z^+ and z^- are the charge numbers of the ions, and ϵ_r is the relative permittivity (or dielectric constant) of the medium.]

An *ion pair*, the constituent ions of which are in direct contact (and not separated by an intervening solvent or other neutral molecule) is designated as a "tight ion pair" (or "intimate" or "contact ion pair"). A tight *ion pair* of X^+ and Y^- is symbolically represented as X^+Y^-.

By contrast, an *ion pair* whose constituent ions are separated by one or several solvents or other neutral molecules is described as a "loose ion pair," symbolically represented as $X^+\|Y^-$. The members of a loose *ion pair* can readily interchange with other free or loosely paired ions in the solution. This interchange may be detectable (e.g., by isotopic labeling) and thus affords an experimental distinction between tight and loose *ion pairs*.

A further conceptual distinction has sometimes been made between two types of loose *ion pairs*. In "solvent-shared *ion pairs*" the ionic constituents of the pair are separated by only a single solvent molecule, whereas in "solvent-separated ion pairs" more than one solvent's molecule intervenes. However, the term "solvent-separated ion pair" must be used and interpreted with care since it has also widely been used as a less specific term for "loose" *ion pair*.

ionizing radiation   Any *radiation* consisting of directly or indirectly ionizing particles or a mixture of both, or photons with *energy* higher than the energy of photons of ultraviolet light or a mixture of both such particles and photons.

irradiance, E   *Radiant power* received by a surface divided by the area of that surface. For collimated beams this quantity is sometimes called intensity and given the symbol I. See also *photon irradiance*.

irradiation   Exposure to *ionizing radiation*.

isomer   One of several species (or *molecular entities*) that have the same atomic composition (molecular formula) but different *line formulae* or different *stereochemical formulae* and hence different physical and/or chemical properties.

isomorphic   Two sets are isomorphic if there is an *isomorphism* between them. Two graphs, G' and G, are *isomorphic* if their nodes can be labeled with the numbers $1, 2, \ldots, p$ such that whenever v_i is adjacent to v_j in G, $\dot{v}_i$ and $\dot{v}_j$ are adjacent in G', $1 \leq i, j \leq p, i \neq j$.

isomorphism   (**1**) A *bijection* between two sets which preserves all structure shared by the sets (e.g., group structure).
(**2**) A morphism between objects of a *category* which is both *surjective* and *injective*.

isotope effect   The effect on the rate or equilibrium constant of two reactions that differ only in the isotopic composition of one or more of their otherwise chemically identical components is referred to as a *kinetic isotope effect* or a *thermodynamic* (or *equilibrium*) *isotope effect*, respectively.

isotopically labeled   Describing a mixture of an *isotopically unmodified* compound with one or more analogous *isotopically substituted* compound(s).

isotopically modified   Describing a compound that has a macroscopic composition such that the isotopic ratio of *nuclides* for at least one element deviates measurably from that occurring in nature. It is either an *isotopically substituted* compound or an *isotopically labeled* compound.

isotopically substituted   Describing a compound that has a composition such that essentially all the molecules of the compound have only the indicated *nuclide(s)* at each designated position. For all other positions, the absence of nuclide indication means that the nuclide composition is the natural one.

isotopically unmodified   Describing a compound that has a macroscopic composition such that its constituent *nuclides* are present in the proportions occurring in nature.

isotopologue   A molecular entity that differs only in isotopic composition (number of isotopic substitutions), e.g., CH_4 CH_3D, CH_2D_2.

isotopomer   *Isomers* having the same number of each isotopic atom but differing in their positions. The term is a contraction of "isotopic isomer." *Isotopomers* can be either constitutional isomers (e.g., $CH_2DCH = 0$ and $CH_3CD = 0$) or isotopic stereoisomers (e.g., (R)- and $(S) - CH_3CHDOH$ or (Z)- and $(E) - CH_3CH = CHD$).

isotropic (sub)manifold (in a symplectic manifold $[P, \omega]$)   A submanifold $M \subset P$ of a symplectic manifold (P, ω) such that at any point $p \in M$ the tangent space contains its symplectic polar, i.e.,

$$(T_pM)^\circ \subset T_pM.$$

The dimension of an *isotropic manifold* M is at most half of the dimension of P.

isotropy group   Let $\Phi : G \times M \to M$ be a *smooth action* of the *Lie group* G on the *manifold* M. For each $x \in M$ the group

$$G_x = \{g \in G \mid \Phi(g, x) = x\}$$

is called the *isotropy group* of Φ at x.

J

jet bundle   Let $(B, M, \pi; F)$ be a *bundle* and $\Gamma_x(\pi)$ be the set of local sections defined around $x \in M$. The jet space at x is the space $J_x^k B$ of equivalence classes, denoted by $j_x^k \sigma$, in $\Gamma_x(\pi)$, of sections having contact k at x (i.e., two local sections are equivalent if they have the same k-order Taylor polynomial). The union of all such jet spaces

$$J^k B = \bigcup_{x \in M} J_x^k B$$

is a bundle over M (as well as over B and over all $J^h B$ for $h \leq k$), and it is called the k-jet prolongation *of B*. If $(x^\mu; y^i)$ are fibered coordinates of B, then $(x^\mu, y^i, y_\mu^i, \ldots, y_{\mu_1 \ldots \mu_k}^i)$ are fibered coordinates of J_B^k. The coordinates $y_{\mu_1 \ldots \mu_h}^i$ are meant to be symmetric with respect to the lower indices $\mu_1 \ldots \mu_h$ for all $1 < h < k$.

Any section $\sigma(x) = (x^\mu, y^i(x))$ induces a section of $J^k B$ defined by

$$j^k \sigma(x) = (x^\mu, y^i(x), \partial_\mu y^i(x), \ldots, \partial_{\mu_1 \ldots \mu_k} y^i(x))$$

which is called the *k-jet prolongation* of σ.

Any bundle morphism $\Phi : B \to B'$ (projecting over a diffeomorphism $\phi : M \to M'$) induces a *bundle morphism* $J^k \Phi : J^k B \to J^k B'$ defined by

$$j^k \Phi(j_x^k \sigma) = j_{f(x)}^k (\Phi \circ \sigma \circ f^{-1})$$

which is called the *k-jet prolongation* of Φ.

Analogously, any projectable vector field X over B induces a projectable vector field $j^k X$ over $J^k B$ which is called the *k-jet prolongation* of X.

junction point   See *branch point*.

K

Karle-Hauptman method   In x-ray crystallography, the diffraction pattern is essentially the *Fourier transform* of the periodic crystal structure which has both an amplitude and a phase. In practice, the phase information cannot be obtained; hence, one has to invert the *Fourier transform* without the phase and, hence, the algorithm is not unique. J. Karle and H. Hauptman developed a system of inequalities which the amplitudes have to satisfy. Using these inequalities as constraints, one can uniquely determine the inverse *Fourier transform* if the number of atoms in a unit cell is sufficiently small. The inequalities are based on the fact that the electron density is a positive function of which the diffraction is the *Fourier transform* (*cf.* J. Karle, *J. Chem. Inf. Comput. Sci.*, 34, 381, 1994).

killing field   On a *Riemannian manifold* (M, g) a *vector field* X such that the *Lie derivative* $L_X g = 0$.

killing vector   Over a pseudo-*Riemannian manifold* (M, g), a *vector* field $\xi = \xi^\mu(x)\, \partial_\mu \in \mathfrak{X}(M)$ such that the *Lie derivative* $\$_\xi g$ vanishes. If $\{^\alpha{}_{\beta\mu}\}_g$ are the *Christoffel symbols* of the metric g, ξ is a *killing vector* if and only if it satisfies the *killing equation*:

$$\nabla_\mu \xi_\nu + \nabla_\nu \xi_\mu = 0$$

where $\xi_\nu = g_{\lambda\nu}\xi^\lambda$ and $\nabla_\mu \xi_\nu = \partial_\mu \xi_\nu - \{^\lambda{}_{\nu\mu}\}_g \xi_\lambda$.
Accordingly, *killing vectors* are infinitesimal generators of *isometries* of g. A killing vector is uniquely determined once one specifies its value $\xi^\mu(x_0)$ at a point $x_0 \in M$ together with its derivatives $\partial_\nu \xi^\mu(x_0)$. (In fact, by deriving of the killing equation one obtains a Cauchy problem which uniquely determines the component functions ξ^μ.) On a *manifold M* of dimension m one can have at most $m(m + 1)/2$ killing vectors. The *manifolds* with exactly $m(m + 1)/2$ killing vectors are called *maximally symmetric*; e.g., the plane, the sphere, and the hypersphere are maximally symmetric.

kinetic energy, E_k Energy of motion. For a body of mass m, $E_k = mv^2/2$, where v is the speed.

kinetic equation   A balanced reaction equation which describes an elementary step in the kinetic sequence of an overall biochemical reaction among unambiguously identified reactants. There is no constraint on the number of participating species. Equally, a set of such equations which can be combined by any allowed operation or combination of operations to produce kinetics identical to those measured for the overall biochemical reaction.
  Comment: The requirement that the equation be elementary ensures that its order will be the sum of the molecularities of the reactants (otherwise, it will be the reactants' activities). Kinetic sequences here considered correspond to the common word descriptors (sequential, ping-pong, etc.). The requirement for identified species distinguishes among different proteins which catalyze the "same" reactions but at different rates. Since most kinetic sequences consist of more than one elementary step, the equations will usually occur in groups. Since not all reactions are sequential, the definition provides that they can be combined by operators other than summation, such as Boolean operators or coefficients representing the frequency of a particular type of event in a population of simultaneously occurring events.

kinetic equivalence   Referring to two reaction schemes which imply the same *rate law*. For example, consider the two schemes (i.) and (ii.) for the formation of C from A:

(i.) $A \underset{k_{-1}, OH^-}{\overset{k_1, OH^-}{\rightleftharpoons}} B \overset{k_2}{\longrightarrow} C$

providing that B does not accumulate as a *reaction intermediate*.

$$\frac{d[C]}{dt} = \frac{k_1 k_2 [A][OH^-]}{k_2 + k_{-1}[OH^-]} \qquad (1)$$

(ii.) $A \underset{k_{-1}}{\overset{k_1}{\rightleftharpoons}} B \overset{k_2}{\underset{OH^-}{\longrightarrow}} C$

Providing that B does not accumulate as a reaction intermediate:

$$\frac{d[C]}{dt} = \frac{k_1 k_2 [A][OH^-]}{k_{-1} + k_2[OH^-]} \qquad (2)$$

Both equations for $d[C]/dt$ are of the form

$$\frac{d[C]}{dt} = \frac{r[A][OH^-]}{1 + s[OH^-]} \qquad (3)$$

where r and s are constants (sometimes called "coefficients in the rate equation"). The equations are identical in their dependence on concentrations and do not distinguish whether OH^- catalyzes the formation of B, and necessarily also its reversion to A, or is involved in its further transformation to C. The two schemes are therefore kinetically equivalent under conditions to which the stated provisos apply.

kinetic motif   A kinetic sequence and associated constants, represented as a set of balanced reactions and their parameters, conserved over different molecules, reactions, or combinations thereof.

Comment: One of several different applications of graph theory to biochemistry is to represent such *kinetic motifs*. Notice there is no constraint on the number of species in the reaction equation. The definition may later be expanded to include noninitial studies. The words used to characterize qualitatively *enzyme* kinetics (ping-pong, ordered) are useful but incomplete. In practice, recognition of the values of the constants will clearly be within tolerances to minimize experimental variation. See also *chemical, dynamical, functional, mechanistic, phylogenetic, regulatory, thermodynamic*, and *topological motives*.

kinetic order   The *kinetic order* of a biochemical reaction is the sum of the *molecularities* of the reaction. A reaction is said to be of nth order in a reactant (partial order) if the stoichiometry of that reactant in the reaction equation is n.

Comment: As defined here, the biochemical reaction corresponds to that for an elementary chemical reaction. The nice correspondence among molecularity, stoichiometry, and order is a direct consequence of the restriction to elementary reactions. In practice, the (partial) order is simply that exponent for a molecular species in a kinetic equation that fully accounts for the production or consumption of that species, the overall order being the sum of the partial orders for all species. Thus until a reaction is established to be elementary, one cannot assume that the exponents will be the *stoichiometries*. If the concentration of a reactant is so large compared to the others as to be "effectively infinite," then the reaction becomes zeroth order for that reactant.

kinetic proofreading   A mathematical model based on the kinetics of transcription (from DNA to RNA) or translation (from RNA to protein) which explains how an extremely high accuracy is achieved in these biological processes in the presence of thermal noise. The key idea is that these biological processes utilize free energy in order to obtain high fidelity (*cf.* J.J. Hopfield, *Proc. Natl. Acad. Sci. U.S.A.*, 71, 4135, 1974).

Kolmogorov equation   Let $U \subset \mathbb{R}^n$ open and $u : U \times \mathbb{R} \to \mathbb{R}$. The *Kolmogorov equation* for u is

$$u_t - \sum_{i,j=1}^{n} a^{ij} u_{x_i x_j} - \sum_{i=1}^{n} b^i u_{x_i} = 0.$$

Korteweg-deVries (KdV) equation   Let $U \subset \mathbb{R}^n$ open and $u : U \times \mathbb{R} \to \mathbb{R}$. The *Korteweg-deVries (KdV) equation* is the shallow water wave equation

$$u_t + u u_x + u_{xxx} = 0.$$

L

label A unique identifier for a *node* or *edge* of a graph, network, or a subnetwork (subgraph).

Comment: Thus, compound names are labels for those nodes, and an edge is labeled by giving the *tuple* of nodes which it joins, so (v_i, v_j).

laboratory sample The sample or subsample(s) sent to or received by the laboratory. When the laboratory sample is further prepared (reduced) by subdividing, mixing, grinding, or by combinations of these operations, the result is the test sample. When no preparation of the *laboratory sample* is required, the *laboratory sample* is the test sample. A test portion is removed from the test sample for the performance of the test or for analysis. The *laboratory sample* is the final sample from the point of view of sample collection but it is the initial sample from the point of view of the laboratory. Several laboratory samples may be prepared and sent to different laboratories or to the same laboratory for different purposes. When sent to the same laboratory, the set is generally considered as a single *laboratory sample* and is documented as a single sample.

Ladyshenskaja-Babuška-Brezzi (LBB) condition A sufficient condition for the stability of finite element schemes for *saddle point problems*. It amounts to requiring uniform *inf-sup conditions* for the pair of *finite element spaces* used in a *Galerkin discretization* of the saddle point problem.

Lagrange finite elements A *parametric equivalent* family of *finite elements* giving rise to $H^1(\Omega)$-conforming finite element spaces. Parametric equivalence is based on the following pullback mapping for functions

$$\mathcal{F}_\Phi(u)(\mathbf{x}) := u(\Phi(x)) \; \mathbf{x} \in K.$$

Thus, it is sufficient to specify the finite elements for reference elements. *Lagrangian* finite elements can be distinguished by their polynomial degree $k \in \mathbb{N}$.

First, consider the unit simplex K in $\mathbb{R}^n$. The local space V_K agrees with the space of multivariate polynomials of total degree $\leq k, k \in \mathbb{N}$,

$$V_K := \left\{ \mathbf{x} \mapsto \sum_{\alpha \in \mathbb{N}_0^n, |\alpha| \leq k} a_\alpha x_1^{\alpha_1} \cdots x_a^{\alpha_n}, a_\alpha \in \mathbb{C} \right\}.$$

The *local degrees of freedom* are based on point evaluations

$$\Pi_k := \left\{ \phi_\mathbf{p} : C(K) \to \mathbb{C}, \phi(u) = u(\mathbf{p}), \mathbf{p} \in \mathbb{P} \right\},$$

where

$$\mathbb{P} := \left\{ \frac{1}{k}(l_1, \cdots, l_n)^T, l_i \in \mathbb{N}_0, \sum_{i=1}^n l_i \leq k \right\} \subset K.$$

Obviously, we have

$$\dim V_K = \binom{n+k}{k}.$$

On the unit hyper-cube $K \subset \mathbb{R}^N$, the geometric reference element for quadrilateral and hexahedral meshes, the local spaces V_K are given by polynomials with degree $\leq k$ in each independent variable

$$V_K := \left\{ \mathbf{x} \mapsto \sum_{\alpha \in \mathbb{N}_0^n, \alpha_i \leq k} a_\alpha x_1^{\alpha_1} \cdots x_a^{\alpha_n}, a_\alpha \in \mathbb{C} \right\}.$$

Local *degrees of freedom* are given by point evaluation in the points

$$\mathbb{P} := \left\{ \frac{1}{k}(l_1, \cdots, l_n)^T, l_i \in \mathbb{N}_0, l_i \leq k \right\} \subset K,$$

which means

$$\dim V_K = (k+1)^n.$$

Lagrangian The *Lagrangian* formulation of mechanics is based on the variational principle of Hamilton. To describe a mechanical system one chooses a configuration space Q with coordinates q^i, $i = 1, \ldots, n$. Then one introduces the *Lagrangian* function $L = K - V$, where K

is the kinetic and V the potential energy. The variational principle states

$$\delta \int_a^b L(q^i, \dot{q}^i, t)dt = 0$$

which leads to the *Euler-Lagrange equations* of motion

$$\frac{d}{dt}\frac{\partial L}{\partial \dot{q}^i} - \frac{\partial L}{\partial q^i} = 0, \quad i = 1, \dots, n.$$

See also *Lagrangian system*.

Lagrangian (sub)manifold (in a symplectic manifold $[P, \omega]$)   A submanifold $M \subset P$ of a symplectic *manifold* (P, ω) which is both *isotropic* and coisotropic, i.e., such that

$$T_p M = (T_p M)^o.$$

The dimension of a *Lagrangian manifold* M is exactly half of the dimension of P.

Lagrangian symmetry   A transformation leaving a *Lagrangian system* invariant.

For example, let (M, L) be a *time-independent Lagrangian system*; a transformation $\Phi : M \to M$, locally given by $\Phi(x) = x'$, can be lifted to the *tangent* space $T\Phi : TM \to TM$, locally given by $T\Phi(x, v) = (x', v')$. It is a *Lagrangian symmetry* if the following identity is satisfied

$$L(x', v') = L(x, v).$$

In field theory, a *bundle morphism* $\Phi : \mathcal{B} \to \mathcal{B}$, which lifts to the jet prolongation $j^k\Phi : J^k\mathcal{B} \to J^k\mathcal{B}$, such that

$$(j^k\Phi)^*L = L.$$

Lagrangian system   A *dynamical system* which is defined by the *action functional* induced by a *Lagrangian*.

A *time-independent Lagrangian system* is a pair (Q, L) where Q is a *manifold*, called the *configuration space*, and $L : TQ \to \mathbb{R}$ is a *function* on the *tangent* space of Q. The *function* L is called the *Lagrangian* of the system.

A *Lagrangian field theory* is a pair $(\mathcal{B}, L)$ where $\mathcal{B} = (B, M, p; F)$ is a *fiber bundle*,

called the *configuration bundle*, and L is a horizontal m-form over the k-jet prolongation $J^k\mathcal{B}$ [m being the dimension of the base manifold $m = \dim(M)$]. The *dynamical system* is induced by the *action functional* on the sections σ of the configuration bundle:

$$A_D(\sigma) = \int_D (j^k\sigma)^*L$$

where $(j^k\sigma)^*L$ denotes evaluation of the Lagrangian L along the (jet prolongation) of the configuration σ. See also *action functional*.

lamp   A source of incoherent radiation.

language   A *language* $\mathcal{L} = (\Sigma, \mathcal{N}, c, \Pi)$ can be described either as a *tuple* of the language's alphabet (Σ), the set of nonterminals ($\mathcal{N}$), the symbol denoting a construct in the language (c), and the set of grammar rules (Π); or equally, the nonempty countable set of allowed constructs $\mathcal{C}, \mathcal{C} \in \Sigma^*$, where Σ^* is the set of all possible constructs formed over the alphabet Σ.

Comment: Programs (for example, databases and queries upon them) are necessarily expressed in some arbitrary language.

Laplace equation   Let $U \subset \mathbb{R}^n$ open and $u : U \to \mathbb{R}$. The *Laplace equation* for u is

$$\Delta u = \sum_{i=1}^n u_{x_i x_i} = 0.$$

Lax-Milgram lemma   This lemma asserts the existence and uniqueness of solutions of variational problems with *elliptic* sesquilinear forms. It is a particular case of more general results for sesquilinear forms satisfying *inf-sup conditions*.

left action (of a group on a space X)   A map $\lambda : G \times X \to X$ such that:

(i.) $\lambda(e, x) = x$;

(ii.) $\lambda(g_1 \cdot g_2, x) = \lambda(g_1, \lambda(g_2, x))$;

where G is a group, e its neutral element, $\cdot$ the product operation in G, and X a topological space. The maps $\lambda_g : X \to X$ defined by $\lambda_g(x) = \lambda(g, x)$ are required to be *homeomorphisms*. A map $\hat{\lambda} : G \to \text{Hom}(X)$ defined by $\hat{\lambda}(g) = \lambda_g$ is thence associated to a left action λ. The *orbit* of a point $x \in X$ is the subset $o_x = \{x' \in X : \exists g \in G, \lambda(g, x) = x'\}$. The action is *transitive* if $o_x = X$ (for any $x \in X$).

The action is *free* if it has no fixed points, i.e., if there exists an element $x \in X$ such that $\lambda(g, x) = x$, then one has $g = e$. Example: *Rotations* in $\mathbb{R}^3$ are not free and not transitive; *translations* are free and transitive.

If X has a further structure, one usually requires λ to preserve the structure. For example, if X is a *topological space*, λ_g are required to be continuous for any $g \in G$.

If $X = V$ is a *vector space*, λ_g are required to be linear for any $g \in G$ and in this case the left action is also called a *representation* of G on V.

If $X = M$ is a *manifold*, λ_g are required to be *diffeomorphisms* for any $g \in G$.

If λ is a left action, then one can define a *right action* by setting $\rho(x, g) = \lambda(g^{-1}, x)$.

left invariance   The property that an object on a *manifold M* is invariant with respect to a *left action* of a group on M. For example, a *vector field* $X \in \mathfrak{X}(M)$ is left invariant with respect to the left action $\lambda_g : M \to M$ if and only if:

$$T_x \lambda_g X(x) = X(g \cdot x).$$

left translations (on a group G)   The *left action* of G onto itself defined by $\lambda(g, h) = \lambda_g(h) = g \cdot h$. Notice that, if G is a *Lie group*, $\lambda_g \in \mathrm{Diff}(G)$ is a *diffeomorphism* but not a *homomorphism* of the group structure. See also *right translations* and *adjoint representations*.

Legendre transformation   Given a *Lagrangian* $L : TQ \to \mathbb{R}$, the *Legendre transformation* $\mathbf{F}L : TQ \to T^*Q$ is defined in local coordinates $(q^i, \dot{q}^i)$ of TQ by

$$\mathbf{F}L(q^i, \dot{q}^i) = (q^i, p_i), \quad \text{where } p_i = \frac{\partial L}{\partial \dot{q}^i}.$$

This gives an equivalence between the *Euler-Lagrange equations* and the *Hamilton equations* of motion.

length   In the context of graph theory, the *length* of a path or cycle is the number of *edges* in the particular subgraph.

Comment: Lengths are an instance of the more general class of measures (such as string length, absolute value of a number, etc.).

Levi-Civita connection   The symmetric (linear) connection $\Gamma^\alpha_{\beta\mu}$ on a (pseudo)-*Riemannian manifold* (M, g) uniquely defined by the requirement that the metric g is parallel, i.e., $\nabla g = 0$. Equivalently, it is the only connection $\Gamma^\alpha_{\beta\mu}$ such that

(i.) it is symmetric, i.e., torsionless, i.e., $\Gamma^\alpha_{\beta\mu} = \Gamma^\alpha_{\mu\beta}$;

(ii.) it is compatible with the (pseudo)-metric structure, i.e.,

$$\partial_\lambda g_{\mu\nu} - \Gamma^\epsilon_{\mu\lambda} g_{\epsilon\nu} - \Gamma^\epsilon_{\nu\lambda} g_{\mu\epsilon} = 0.$$

The coefficients of the *Levi-Civita connection* are usually denoted by $\Gamma^\alpha_{\beta\mu} = \left\{{\alpha \atop \beta\mu}\right\}_g$ and can be expressed as a function of the (pseudo)-metric tensor and its first derivatives; they are called *Christoffel symbols*.

Lewis acid   A *molecular entity* (and the corresponding *chemical species*) that is an electron-pair acceptor and, therefore, able to react with a *Lewis base* to form a *Lewis adduct*, by sharing the electron pair furnished by the Lewis base.

Lewis adduct   The *adduct* formed between a *Lewis acid* and a *Lewis base*.

Lewis base   A *molecular entity* (and the corresponding *chemical species*) able to provide a pair of electrons and thus capable of *coordination* to a *Lewis acid*, thereby producing a *Lewis adduct*.

Lie algebra   A *vector space L* endowed with an operation $[,] : L \times L \to L$ having the following properties

(i.) $[,]$ is bilinear;

(ii.) $[,]$ is antisymmetric (or skew-symmetric), i.e., $[A, B] = -[B, A]$;

(iii.) $[,]$ satisfies *Jacoby identity* (i.e., $[[A, B], C] + [[B, C], A] + [[C, A], B] = 0$).

The operation $[,]$ is called the *Lie bracket* or *commutator*. Notice that usually *Lie algebras* are not associative. See *Lie group* for the definition of a *Lie algebra of the Lie group G*.

Lie derivative   Let α be an *exterior k-form* and X a *vector field* with *flow* φ_t. The *Lie derivative* of α along X is given by

$$\mathbf{L}_X \alpha = \lim_{t \to 0} \frac{1}{t} [(\varphi_t^* \alpha) - \alpha].$$

Lie group   A group G which has a compatible *manifold* structure such that both the product $\cdot$: $G \times G \to G$ and the inversion $i : G \to G$: $g \mapsto g^{-1}$ are differentiable maps.

Left translations $L_g : G \to G$ defined by $L_g(h) = g \cdot h$ are *diffeomorphisms*. If $v \in T_e G$ is a *tangent* vector in the unit $e \in G$ we can define a *vector* field $\lambda_v(g) = T_e L_g(v)$ which is *left-invariant*, i.e., $T L_g \lambda_v(h) = \lambda_v(g \cdot h)$. Any left-invariant *vector* field on G is obtained in this way. The set $\mathfrak{X}_L(G)$ of left-invariant *vector* fields form a *Lie subalgebra* of the *Lie algebra* of all *vector* fields $\mathfrak{X}(G)$. The *Lie algebra* $\mathfrak{X}_L(G)$ is *isomorphic* to $T_e G$ as a *vector space*, so that it is finite dimensional ($\dim(\mathfrak{X}_L(G)) = \dim T_e G = \dim G$). It is called *the Lie algebra of G*.

Analogously, *right-translations* $R_g : G \to G$ defined by $R_g(h) = h \cdot g$ are *diffeomorphisms*. If $v \in T_e G$ is a *tangent* vector we can define a *vector field* $\rho_v(g) = T_e R_g(v)$ which is right-invariant, i.e., $T R_g \rho_v(h) = \rho_v(h \cdot g)$. Any right-invariant *vector* field on G is obtained in this way. The set $\mathfrak{X}_R(G)$ of right-invariant *vector* fields form a Lie subalgebra of the *Lie algebra* of *vector* fields $\mathfrak{X}(G)$. The *Lie algebra* $\mathfrak{X}_R(G)$ is canonically *isomorphic* to $\mathfrak{X}_L(G)$ as a *vector space*, so that it is also finite dimensional ($\dim(\mathfrak{X}_R(G)) = \dim G$). Clearly, if the group G is Abelian, then $\mathfrak{X}_L(G)$ and $\mathfrak{X}_R(G)$ coincide.

Lie-Poisson bracket   Let **g** be a *Lie algebra*, **g*** its dual space and $< \ , \ >: \mathbf{g}^* \times \mathbf{g} \to \mathbb{R}$ the natural pairing, $< \mu, \xi > = \mu(\xi)$. The *Lie-Poisson bracket* of any $F, G : \mathbf{g}^* \to \mathbb{R}$ is defined by

$$\{F, G\}_\pm(\mu) = \pm \langle \mu, \left[\frac{\delta F}{\delta \mu}, \frac{\delta G}{\delta \mu} \right] \rangle$$

where $\frac{\delta F}{\delta \mu} \in \mathbf{g}$ is the functional derivative of F at μ defined by

$$\lim_{t \to 0} \frac{1}{t} [F(\mu + tv) - F(\mu)] = \langle v, \frac{\delta F}{\delta \mu} \rangle.$$

With the *Lie-Poisson bracket*, **g*** is a *Poisson manifold*.

lifetime (mean lifetime), τ   The lifetime of a *chemical species* which decays in a first-order process is the time needed for a concentration of this species to decrease to $1/e$ of its original

value. Statistically, it represents the mean life expectancy of an excited species. In a reacting system in which the decrease in concentration of a particular chemical species is governed by a first-order *rate law*, it is equal to the reciprocal of the sum of the (pseudo)unimolecular rate constants of all processes which cause the decay. When the term is used for processes which are not first order, the lifetime depends on the initial concentration of the species, or of a quencher, and should be called *apparent* lifetime instead.

limiting current   The limiting value of a *faradaic current* that is approached as the rate of the charge-transfer process is increased by varying the potential. It is independent of the *applied potential* over a finite range, and is usually evaluated by subtracting the appropriate *residual current* from the measured total current. A limiting current may have the character of an *adsorption, catalytic, diffusion,* or *kinetic current,* and may include a *migration current.*

line formula   A two-dimensional representation of *molecular entities* in which atoms are shown joined by lines representing single or multiple bonds, without any indication or implication concerning the spatial direction of bonds.

linear   A *vector space* is often called a *linear space*. A map $T : V \to W$ from a linear space V into a linear space W is called *linear* or a *linear transformation* or a *linear operator* if $T(x + y) = T(x) + T(y)$ and $T(\alpha x) = \alpha T(x)$ for all $x, y \in V, \alpha \in \mathbb{C}$.

linear chain   A *chain* with no *branch points* intermediate between the boundary units.

linear functional   A *linear* scalar valued function $f : V \to \mathbb{C}$ on a *vector space V*. Sometimes continuity of f is also assumed.

linear group (of a *vector space V*)   The group of *automorphisms* of V (which is a group with respect to composition). It is denoted by GL(V).

Also the matrix group of all invertible (finite) matrices. It can be noncanonically identified with GL($\mathbb{R}^m$), and it is denoted by GL($m, \mathbb{R}$). If $\dim(V) = m$, then there exists a group isomorphism between GL(V) and GL($m, \mathbb{R}$) which is induced by the choice of a basis of V; both GL(V) and GL($m, \mathbb{R}$) are *Lie groups*.

linear macromolecule   A *macromolecule*, the structure of which essentially comprises the multiple repetition in linear sequence of units derived, actually or conceptually, from molecules of low relative molecular mass.

linear operator   See *linear*.

linear transformation   See *linear*.

Liouville's equation   Let $U \subset \mathbb{R}^n$ open and $u : U \times \mathbb{R} \to \mathbb{R}$. The *Liouville equation* for u is

$$u_t - \sum_{i=1}^{n} (b^i u)_{x_i} = 0.$$

liquid-crystal transitions   A liquid crystal is a molecular crystal with properties that are both solid- and liquid-like. Liquid crystals are composed predominantly of rod-like or disk-like molecules, that can exhibit one or more different, ordered fluid phases as well as the isotropic fluid; the translational order is wholly or partially destroyed but a considerable degree of orientational order is retained on passing from the crystalline to the liquid phase in a *mesomorphic transition*.

(**1**) Transition to a *nematic phase*. A mesomorphic transition that occurs when a molecular crystal is heated to form a nematic phase in which the mean direction of the molecules is parallel or antiparallel to an axis known as the director.

(**2**) Transition to a *cholesteric phase*. A mesomorphic transition that occurs when a molecular crystal is heated to form a cholesteric phase in which there is simply a spiraling of the local orientational order perpendicular to the long axes of the molecules.

(**3**) Transition to a *smectic state*. A mesomorphic transition that occurs when a molecular crystal is heated to yield a smectic state in which there is a one-dimensional density wave which produces very soft/disordered layers.

list   An unordered collection of elements, that may include duplicates. An enumerated list is delimited by brackets ([x]).

Comment: Notice the definitions of *set*, *bag*, and *list* progressively release constraints on the elements in the various collections. The elements of sets and bags have some relationship to each

other (for example, they satisfy a generating relation like $y = x^2$ or are all employees in a company), but either may or may not have repeated elements. The elements of a list need not have any relationship to each other. Notice further that none of these definitions requires that the members of the collection be ordered in some way. If they were, one would speak of an *ordered set*, *bag*, or *list*. For example, one might sort the elements of a set alphabetically or in UNIX sort order. See also *bag, sequence, set*, and *tuple*.

load vector   The right-hand side of a variational problem posed over the *Banach space* V is an element of the dual space V'. When discretization by means of a conforming *finite element space* V_h is performed, f has to be evaluated for the *nodal basis functions* $b_i, i = 1, \cdots, N := \dim V_h$, of V_h. The resulting vector $(f(b_i))_{i=1}^{N}$ has been dubbed *load vector* in calculations of linear elasticity.

logistic equation   A nonlinear equation with a $x(1 - x)$ term. It was first motivated from modeling biological population growth. In the context of ordinary differential equations, the equation, $dx/dt = ax(1 - x)$ can be solved. In the context of difference equations, $x_{n+1} = ax_n(1 - x_n)$. This equation exhibits a wide range of interesting nonlinear phenomena: bifurcation and periodic doubling to chaos.

London forces   Attractive forces between apolar molecules, due to their mutual polarizability. They are also components of the forces between polar molecules. Also called "dispersion forces."

loop (based at $x_0 \in X$)   A curve $\gamma : I \to X$ such that $\gamma(0) = \gamma(1) = x_0$ and $I = [0, 1]$. The set of all loops of X based at $x_0 \in X$ can be endowed with a group product. Let $\lambda, \gamma : I \to X$ be two loops; the product $\lambda * \gamma : I \to X$ is the following loop

$$\lambda * \gamma(t) = \begin{cases} \gamma(2t) & 0 \le t \le 1/2 \\ \lambda(2t - 1) & 1/2 \le t \le 1. \end{cases}$$

The group structure is not commutative. It is compatible with the *homotopy* equivalence relation so that it induces the group structure of the *homotopy group* $\pi_0(X, x_0)$ of X (based at x_0).

Let us now consider the homotopy groups $\pi_0(X, x_0)$ and $\pi_0(X, x_1)$ at two different points of X and a path $\gamma : I \to X$ connecting the two points, i.e., $\gamma(0) = x_0$ and $\gamma(1) = x_1$. We can define a group *isomorphism i* between $\pi_0(X, x_0)$ and $\pi_0(X, x_1)$ given by

$$i(\lambda) = \gamma^{-1} * \lambda * \gamma \in \pi_0(X, x_0) \qquad \lambda \in \pi_0(X, x_1)$$

where γ^{-1} is the path $\gamma^{-1}(t) = \gamma(1 - t)$.

Lorentz group   The special *orthogonal group* $SO(1, m - 1)$ of *isometries* of $\mathbb{R}^m$ with the standard indefinite metric η of signature $(1, m-1)$. Let us choose an orthonormal basis E_i in $\mathbb{R}^m$, the metric is expressed by $\eta = \eta_{ij} E^i \otimes E^j$ (where E^i is the dual basis). The matrix η_{ij} is defined by $\eta_{00} = 1$, $\eta_{ii} = -1$ when $i > 0$ and $\eta_{ij} = 0$ when $i \neq j$. An endomorphism $\alpha : \mathbb{R}^m \to \mathbb{R}^m$ is an element of the *Lorentz group* if and only if

$$A_i^a \, \eta_{ab} \, A_j^b = \eta_{ij} \qquad\qquad \alpha(E_i) = A_i^j \, E_j.$$

Lorentzian manifold   A pair (M, g) formed by a *manifold M* and a Lorentzian metric g on M, i.e., a pseudo-Riemannian metric of signature either $(1, m - 1)$ or $(m - 1, 1)$.

luminescence   Spontaneous emission of radiation from an electronically or vibrationally excited species not in thermal equilibrium with its environment.

lyate ion   The *anion* produced by *hydron* removal from a solvent molecule. For example, the hydroxide ion is the *lyate ion* of water.

M

macromolecule (polymer molecule)   A molecule of high relative molecular mass, the structure of which essentially comprises the multiple repetition of units derived, actually or conceptually, from molecules of low relative molecular mass.

Notes: (**1**) In many cases, especially for synthetic polymers, a molecule can be regarded as having a high relative molecular mass if the addition or removal of one or a few of the units has a negligible effect on the molecular properties. This statement fails in the case of certain *macromolecules* for which the properties may be critically dependent on fine details of the molecular structure.

(**2**) If a part of the whole of the molecule has a high relative molecular mass and essentially comprises the multiple repetition of units derived, actually or conceptually, from molecules of low relative molecular mass, it may be described as either *macromolecular* or polymeric, or by polymer used adjectivally.

manifold   A (usually connected) *topological space* M such that there exists an *atlas* $\{(U_\alpha, \varphi_\alpha)\}_{\alpha \in I}$, i.e., a collection such that:

(i.) $\{U_\alpha\}_{\alpha \in I}$ is an open covering of M;

(ii.) $\varphi_\alpha : U_\alpha \to W_\alpha$ is a local *homeomorphism* onto $W_\alpha = \varphi_\alpha(U_\alpha) \subset \mathbb{R}^m$;

(iii.) $\varphi_\beta \circ \varphi_\alpha^{-1} : W_\alpha \to W_\beta$ are local *homeomorphisms* of class C^k.

The maps $\varphi_\beta \circ \varphi_\alpha^{-1} : W_\alpha \to W_\beta$ are called *transition functions*. The integer m is the *dimension* of M. A *manifold* M is usually assumed to be *paracompact*.

If $k = 0$, M is called a *topological (real) manifold*; if $k = \infty$, it is called a *differentiable (real) manifold*. If the *transition functions* are analytical, then M is called an *analytical (real) manifold*. If *charts* $(U_\alpha, \varphi_\alpha)$ are valued in $\mathbb{C}^m$ and *transition functions* are, *holomorphic*, then the *manifold* is called a *complex manifold*.

For example, the sphere $S^2 \subset \mathbb{R}^3$ ($x^2 + y^2 + z^2 = 1$) is both a real and a complex *manifold*. The real structure is given by the real charts

$$\varphi_N(x, y, z) = \left(\frac{x}{1-z}, \frac{y}{1-z} \right), \quad z \neq 1 \text{ and}$$

$$\varphi_S(x, y, z) = \left(\frac{x}{1+z}, \frac{y}{1+z} \right), \quad z \neq -1$$

while the complex atlas is given by

$$\varphi_N(x, y, z) = \frac{x+iy}{1-z}, \quad z \neq 1 \text{ and}$$

$$\varphi_S(x, y, z) = \frac{x-iy}{1+z}, \quad z \neq -1.$$

mapping   See *function*.

mass, m   Base quantity in the system of quantities upon which SI is based.

mass lumping   Approximation of the *mass matrix* by a diagonal matrix. This is usually achieved by replacing the $L^2(\Omega)$-inner product with a mesh-dependent inner product $(., .)_h$ on a *finite element space* V_h that is based on numerical quadrature. A prominent example is supplied by linear *Lagrangian finite elements* and *vertex* based quadrature: on a triangular *mesh* Ω_h it boils down to

$$(u_h, v_h)_{L^2(\Omega)} \approx \sum_{T \in \mathcal{T}_h} \frac{|T|}{3} \sum_{i=1}^{3} u_h(\mathbf{a}_i^T) \bar{v}_h(\mathbf{a}_i^T)$$

$$\times u_h, v_h \in V_h,$$

where $\mathbf{a}_1^T, \mathbf{a}_2^T, \mathbf{a}_3^T$ stand for the *vertices* of the triangle T. As the *degrees of freedom* for V_h are associated with *vertices* of the mesh, the diagonal structure of the resulting "lumped" *mass matrix* is evident.

mass matrix   The matrix arising from a finite element discretization of the $L^2(\Omega)$-inner product ($\Omega \subset \mathbb{R}^n$ a bounded domain): if $\psi_1 \cdots , \psi_N$ stands for the *nodal basis* of the *finite element space* $V_h \subset L^2(\Omega)$, then the corresponding *mass matrix* is given by $((\psi_i, \psi_j)_{L^2(\Omega)})_{i,j=1}^{N}$. The term *mass matrix* has its origin in the finite element analysis of problems of linear elasticity.

1-58488-050-3/01/$0.00+$.50

mass spectrometer   An instrument in which beams of ions are separated (analyzed) according to the quotient mass/charge, and in which the ions are measured electrically. This term should also be used when a scintillation detector is employed.

matrix   (1) (in analysis) A rectangular array of scalars. A matrix with n rows and m columns represents a linear operator from an m dimensional vector space to an n dimensional vector space.
(2) (in analytical chemistry) The components of the sample other than the analyte.

matrix effect   (1) (in analytical chemistry) The combined effect of all components of the sample other than the analyte on the measurement of the quantity. If a specific component can be identified as causing an effect then this is referred to as interference. See *matrix*.
(2) (in surface analysis) Effects which cause changes in Auger-electron, photoelectron, secondary ion yield, or scattered ion intensity, the energy or shape of the signal of an element in any environment as compared to these quantities in a pure element.
(a) Chemical matrix effects: changes in the chemical composition of the solid which affect the signals as described above.
(b) Physical matrix effects: topographical and/or crystalline properties which affect the signal as described above.

maximal common subgraph   Two graphs, G_1 and G_2, have a *maximal common subgraph* if there exists a graph, G', which is the largest subgraph common to G_1 and G_2.

maximal element (in a partially ordered set $[A, \rhd]$)   Referring to an element $m \in A$ such that there is no element $a \in A$ except $a = m$ such that $a \rhd m$. See *ordering* and *minimum*.

maximum (in a partially ordered set $[A, \rhd]$) Referring to an element $m \in A$ such that $\forall a \in A$, $m \rhd a$. See *ordering*.

Maxwell's equations   In electrodynamics, for an electric field $\mathbf{E}(x, t)$ and magnetic field $\mathbf{B}(x, t)$ Maxwell's equations in the vacuum are
$$\begin{cases} \mathbf{E}_t = curl\ \mathbf{B} \\ \mathbf{B}_t = -curl\ \mathbf{E} \\ div\ \mathbf{B} = div\ \mathbf{E} = 0. \end{cases}$$

Maxwell-Vlasov equations   Let $f(x, v, t)$ denote the plasma density and $\mathbf{E}(x, t), \mathbf{B}(x, t)$ the electric and magnetic fields, respectively. The equations for a collisionless plasma of a single species of charged particles with charge e and mass m are described by the *Maxwell-Vlasov equations*
$$\begin{cases} \frac{\partial f}{\partial t} + v \cdot \frac{\partial f}{\partial x} + \frac{e}{m}(\mathbf{E} + \frac{1}{c}v \times \mathbf{B}) \cdot \frac{\partial f}{\partial v} = 0, \\ \frac{1}{c}\frac{\partial \mathbf{B}}{\partial t} = -curl\ \mathbf{E}, \\ \frac{1}{c}\frac{\partial \mathbf{E}}{\partial t} = -curl\ \mathbf{B} - \frac{1}{c}\mathbf{j}_f, \\ div\ \mathbf{E} = \rho_f \ \text{ and } \ div\mathbf{B} = 0 \end{cases}$$
where the current of f is given by $\mathbf{j}_f = e \int vf(x, v, t)dv$, and the charge density by $\rho_f = e \int f(x, v, t)dv$.

measurable set   A *measure* on a set S is frequently defined only on certain subsets of S, forming a σ algebra (closed under countable unions and differences of its members). Sets in this σ algebra are called *measurable sets*.

measure   (1) A numerical determination of size. For example, the cardinality, if x is a set; or the length, if x is a sequence; or the absolute value, if x is a number.
(2) A nonnegative, real valued function μ defined on certain subsets of a set S and satisfying

(i.) $\mu(\Phi) = 0$

(ii.) $\mu(\cup U_n) = \sum \mu(U_n)$ when $\{U_n\}$ is a pairwise disjoint, countable sequence of sets.

mechanism (of a reaction)   A detailed description of the process leading from the *reactants* to the products of a reaction, including a characterization as complete as possible of the composition, structure, energy, and other properties of *reaction intermediates*, *products*, and *transition states*. An acceptable mechanism of a specified reaction (and there may be a number of such alternative mechanisms not excluded by the evidence) must be consistent with the reaction *stoichiometry*, the *rate law*, and with all other available experimental data, such as the *stereochemical* course of the reaction. Inferences concerning the electronic motions which dynamically interconvert successive species along the *reaction path* (as represented by curved arrows, for example) are often included in the description of a *mechanism*.

It should be noted that for many reactions all this information is not available and the suggested *mechanism* is based on incomplete experimental data. It is not appropriate to use the term *mechanism* to describe a statement of the probable sequence in a set of *stepwise reactions*. That should be referred to as a *reaction sequence*, and not a *mechanism*.

mechanistic equation   A balanced elementary reaction equation, bimolecular or less on one side, which describes the ligand binding, organic chemical, and conformational rearrangement reactions which occur on an unambiguously identified entity, whose net effect in the biochemical reaction is catalytic, to produce an overall biochemical reaction. Equally, a set of such reaction equations which can be combined by summation or logical operators to produce the overall biochemical reaction.

Comment: The requirement that the equation be elementary ensures that its order will be the sum of the molecularities of the reactants. All types of interactions are included. Since the aim of the equations is to describe interactions at the catalytic species, it follows that that molecule will not appear in these equations as a catalyst. The requirement for clearly identified species distinguishes among different catalysts carrying out the "same" reactions at different rates. The definition provides that they can be combined by summation or logical operators to represent alternative paths through the space of possible reactions, depending on the consequences of prior reactions in the set.

mechanistic motif   A repeated pattern over *mechanistic equations*.

Comment: See also *biochemical, chemical, dynamical, functional, kinetic, phylogenetic, regulatory, thermodynamic*, and *topological motives*.

medium   The phase (and composition of the phase) in which *chemical species* and their reactions are studied in a particular investigation.

meiosis   The reductive cell division which results in daughter cells containing one copy of each of the *chromosomes* of the parent. The entire meiotic process involves two separate divisions (meiosis I and meiosis II). The first division is a true reductive division with the chromosome number being halved, whereas the second division resembles mitosis in many ways. Thus, a diploid parental cell will give rise to haploid daughter cells (*cf. ploidy*).

memory   A *memory* is a storage device of fixed half-life for data and programs that permit any combination of reading from and writing to it.

mesh width   The *mesh width* of a *triangulation* Ω_h is the maximal *diameter* of its cells.

mesomeric effect   The effect (on reaction rates, ionization, equilibria, etc.) attributed to a substituent due to overlap of its p- or π-orbitals with the p- or π-orbits of the rest of the *molecular entity. Delocalization* is thereby introduced or extended, and electronic charge may flow to or from the substituent. The effect is symbolized by M.

Strictly understood, the *mesomeric effect* operates in the ground electronic state of the *molecule*. When the *molecule* undergoes electronic excitation or its energy is increased on the way to the *transition state* of a *chemical reaction, the mesomeric effect* may be enhanced by the *electromeric effect*, but this term is not much used, and the *mesomeric* and electronic effects tend to be subsumed in the term *resonance effect* of a *substituent*.

mesomorphic transition   A transition that occurs between a fully ordered crystalline solid and an isotropic liquid. *Mesomorphic transitions* can occur

(i.)   from a crystal to a liquid crystal,

(ii.)   from a liquid crystal to another liquid crystal, and

(iii.)   from a liquid crystal to an isotropic liquid.

mesopause (in atmospheric chemistry)   That region of the atmosphere between the *mesosphere* and the *thermosphere* at which the temperature is a minimum.

mesosphere   That region of the atmosphere which lies above the *stratopause* (about 4752 km) and below the *mesopause* (about 8090 km) and in which temperature decreases with increasing height; this is the region in which the lowest temperatures of the atmosphere occur.

metabolite   A molecule which participates noncatalytically in metabolism and is not an inorganic ion or element.

Comment: This seems a bit circular, because it depends on being able to recognize what metabolism is. Metabolites tend to be relatively small (usually less than 1000 daltons molecular weight) and structurally simple. They can be *monomers* or *polymers* (for example, fatty acids), but are not usually considered to be *macromolecules* or their smaller *polymers* (oligonucleotides, peptides). As should be obvious, the notion of a *metabolite* is rather elastic.

metric   A metric on a set M is a map d : $M \times M \to$ R satisfying for all $x, y, z \in M$,

(i.) $d(x, y) \geq 0$, equal 0 iff $x = y$, positive definiteness,

(ii.) $d(x, y) = d(y, x)$, symmetry,

(iii.) $d(x, z) \leq d(x, y) + d(y, z)$, triangle inequality.

A metric space is a pair (M, d).

metric space   See *metric*.

Michaelis-Menten kinetics   The mathematical model for enzyme kinetics based on the law of mass action. The resulting equations are nonlinear. However, due to the peculiar nature of the enzymatic reaction in which the concentration of the enzyme is usually significantly greater than that of the substrate, the system of nonlinear ordinary differential equations can be treated by singular perturbation. This treatment gives the steady-state approximation well known in biochemical literature.

microscopic reversibility   The continuous reaction of sinistralateral and dextralateral sets of coreactants, such that no net change in the concentrations of coreactants occurs over measurable time.

Comment: Once equilibrium is attained, any reaction will continue in both directions, shifting the reaction away from and back toward equilibrium from instant to instant, but by infinitesimally small changes. See also *dextralateral, direction, dynamic equilibrium, formal reaction equation, product, rate constant, reversibility, sinistralateral,* and *substrate*.

migration   (**1**) The (usually *intramolecular*) transfer of an atom or *group* during the course of a *molecular rearrangement*.

(**2**) The movement of a bond to a new position, within the same *molecular entity*, is known as "bond migration."

Allylic rearrangements, e.g.,

$$RCH = CHCH_2X \longrightarrow RCCHCH = CH_2$$
$$\overset{|}{X}$$

exemplify both types of *migration*.

minimal biochemical network   The *minimal biochemical network* is the network

$$N_{B_0}(\{v_{m,1}, v_{m,2}, v_{r,1}\}, \{e(s, v_{m,1}, v_{r,1}),$$

$$e(d, v_{m,2}, v_{r,1})\}, \emptyset, \{l_{m,1}, l_{m,2}, l_{r,1}\}).$$

Comment: This is a network of a single spontaneous reaction, without any known parameters but for which the identities of the reactants (but not their *stoichiometry*, a parameter of the edges) are known.

minimal element (of a partially ordered set $[A, \rhd]$)   An element $m \in A$ such that there is no element $a \in A$ except $a = m$ such that $m \rhd a$. See *ordering* and *minimum*.

minimal surface equation   Let $U \subset \mathbb{R}^n$ open and $u : U \times \mathbb{R} \to \mathbb{R}$. The minimal surface equation for u is

$$div \left(\frac{Du}{(1 + |Du|^2)^{1/2}} \right) = 0,$$

where $Du = D_x u = (u_{x_1}, \ldots, u_{x_n})$ denotes the gradient of u with respect to the spatial variable $x = (x_1, \ldots, x_n)$.

minimum (of a partially ordered set $[A, \rhd]$)   An element $m \in A$ such that $\forall a \in A, a \rhd m$. See *ordering*.

minimum-energy reaction path   The path corresponding to the steepest descent from the *col* of a *potential-energy surface* into the two valleys. The reaction coordinate corresponds to this minimal path. Some workers refer to the minimum-energy reaction path as simply the reaction path but this is not recommended as it leads to confusion.

Minkowski space   The metric *vector space* $(\mathbb{R}^4, \eta)$ with the signature $(1, 3)$.

mixed finite elements   A rather general term for *finite element spaces* meant to approximate *vector fields*. Prominent representatives are

(i.) $\mathbf{H}(\text{div}; \Omega)$-conforming Raviart-Thomas elements required for the discretization of *mixed variational formulations* of second-order elliptic boundary value problems.

(ii.) $\mathbf{H}(\mathbf{curl}; \Omega)$-conforming Nédélec elements (edge elements) used to approximate electromagnetic fields.

(iii.) a variety of schemes for the approximation of velocity fields in computational fluid mechanics.

mixed variational formulation   Consider a second-order elliptic boundary value problem

$$-\text{div}(A\text{grad}u) = f \text{ on } \Omega \subset \mathbb{R}^n, \quad u_{|\partial\Omega} = g_D.$$

Formally its mixed formulation is obtained by introducing the *flux* $\mathbf{j} := A\text{grad}u$ as new unknown, which results in a system of first-order differential equations

$$\mathbf{j} - A\text{grad}u = 0, \quad -\text{div}\mathbf{j} = f.$$

The first equation is tested with $\mathbf{v} \in \mathbf{H}(\text{div}; \Omega)$ and integration by parts is carried out. The second equation is tested with $q \in L^2(\Omega)$. We end up with the variational problem: seek $\mathbf{j} \in \mathbf{H}(\text{div}; \Omega)$, $u \in L^2(\Omega)$ such that

$$\int_\Omega A^{-1}\mathbf{j} \cdot \mathbf{v}d\mathbf{x} + \int_\Omega \text{div}\mathbf{v}ud\mathbf{x} = \int_\Gamma g\mathbf{v} \cdot \mathbf{n}dS$$

$$\forall \mathbf{v} \in \mathbf{H}(\text{div}; \Omega),$$

$$\int_\Omega \text{div}\mathbf{j}qd\mathbf{x} = -\int_\Omega fqd\mathbf{x} \ \forall q \in L^2(\Omega).$$

This is a symmetric *saddle point problem*, which can be discretized by means of $\mathbf{v} \in \mathbf{H}(\text{div}; \Omega)$-conforming *mixed finite elements* for the flux unknown. Solving the second-order elliptic boundary value problem is equivalent to minimizing a convex energy functional on $H_0^1(\Omega)$. Convex analysis teaches that this gives rise to *primal* and *dual* variational problems. The latter leads to the mixed variational formulation.

module (over a ring R)   A set M endowed with an inner binary operation, called the *sum* $+ : M \times M \to M$ and a binary operation called the *product (by a scalar)* $\cdot : R \times M \times M$. The structure $(M, +)$ is a commutative group and the product by scalars satisfies the following axioms:

(i.) $\forall \lambda, \mu \in R, \forall v \in M$: $\lambda(\mu \cdot v) = (\lambda\mu) \cdot v$

(ii.) $\lambda \cdot (v + w) = \lambda \cdot v + \lambda \cdot w$

(iii.) $(\lambda + \mu) \cdot v = \lambda \cdot v + \mu \cdot v$

(iv.) $\mathbb{I}_R \cdot v = v$

If R is a *field* then an R-module is nothing but a *vector space*.

mole   An Avogadro's number of molecules $(\mathcal{N} \approx 6.023 \times 10^{23})$.

molecular dynamics   Based on Newton's law of motion and treating atoms in a *molecule* as classical particles with interaction, molecular dynamics simulate the motion of the entire molecule on a computer. This approach has been successful in studying the dynamics of small molecules but has not yielded a complete picture of the dynamics of biologically important molecules, e.g., proteins. The difficulty is mainly the stiff nature of the many-body system and the limited computational power.

molecular entity   Any constitutionally or isotopically distinct atom, molecule, ion, ion pair, radical, radical ion, complex, conformer, etc., identifiable as a separately distinguishable entity.

Molecular entity is used here as a general term for singular entities, irrespective of their nature, while *chemical species* stands for sets or ensembles of *molecular entities*. Note that the name of a compound may refer to the respective *molecular entity* or to the chemical species; e.g., methane, may mean a single molecule of CH_4 (*molecular entity*) or a molar amount, specified or not (chemical species), participating in a reaction.

The degree of precision necessary to describe a *molecular entity* depends on the context. For example "hydrogen molecule" is an adequate definition of a certain *molecular entity* for some purposes, whereas for others it is necessary to distinguish the electronic state and/or vibrational state and/or nuclear spin, etc. of the hydrogen molecule.

molecularity   The number of reactant *molecular entities* that are involved in the microscopic chemical event constituting an *elementary reaction*. (For reactions in solution this number is always taken to exclude *molecular entities* that form part of the *medium* and which are involved solely by virtue of their solvation of solutes.)   A reaction with a molecularity of one is called "unimolecular," one with a molecularity of two "bimolecular," and of three "termolecular."

moment of a force, M, about a point   The vector product of the radius vector from this point to a point on the line of action of the force and the force, $M = r \times r \times F$.

momentum, p   Vector quantity equal to the product of *mass* and *velocity*.

momentum map   Let a *Lie group G* act on a *Poisson manifold P* and let $\mathbf{g}$ be the *Lie algebra* of G. Any $\xi \in \mathbf{g}$ generates a vector field ξ_P (the infinitesimal generator of the action) on P by

$$\xi_P(x) = \frac{d}{dt}[\exp(t\xi) \cdot x]|_{t=0}.$$

Suppose there is a map $J : \mathbf{g} \to C^\infty(P)$ such that $X_{J(\xi)} = \xi_P$, for all $\xi \in \mathbf{g}$. The map $J : P \to \mathbf{g}^*$ defined by

$$\langle J(x), \xi \rangle = J(\xi)(x) , \quad \xi \in \mathbf{g}, x \in P$$

is called the *momentum map* of the action.

Monge-Ampere equation   Let $U \subset \mathbb{R}^n$ be open and $u : U \times \mathbb{R} \to \mathbb{R}$. The *Monge-Ampere equation* for u is

$$det(D^2u) = f,$$

where $Du = D_x u = (u_{x_1}, \ldots, u_{x_n})$ denotes the gradient of u with respect to the spatial variable $x = (x_1, \ldots, x_n)$.

monomer   A substance composed of *monomer molecules*.

monomer molecule   A molecule which can undergo *polymerization* thereby contributing *constitutional units* to the essential structure of a *macromolecule*.

monomeric unit (nomoner unit, mer)   The largest *constitutional unit* contributed by a single *monomer molecule* to the structure of a *macromolecule* or *oligomer molecule*.

Note: The largest constitutional unit contributed by a single *monomer molecule* to the structure of a *macromolecule* or *oligomer molecule* may be described either as *monomeric* or by *monomer* used adjectivally.

monomorphism   An injective morphism between objects of a *category*. For example, a *monomorphism* of *vector spaces* is a linear injective map; a *monomorphism* of groups is an injective *group homomorphism*; a *monomorphism* of *manifolds* is an injective differentiable map. See also *bundle morphisms*.

motif   For any collection of objects $X = \{x_1, x_2, \ldots, x_n\}$, x_i is a *motif* if there exists an $x_j \in X, i \neq j$ such that $f(x_i) = x_j$. If x_i occurs in X at least twice, that is, if $x_i = x_j$ for $1 \leq i, j \leq n, i \neq j$, it is an *exact motif* or an *exact match*.

Comment: The relationship f is a transformation between x_i and x_j. For nucleic acid and protein sequences, the most common motives are the percent identity and percent similarity relationships, which are usually set to some threshold value such that the identity (similarity) of the resulting string is at least that threshold value. When X is a set of graphs, then f is either the relationship of subgraph (for motif) or isomorphism (for exact motif).

For our purposes the term "pattern" is synonymous with *motif*. It is not obvious that the notion of *motif* as used in biology is isomorphic to that of pattern. Definitions of *motif* and pattern have been developed to cover problems arising from tessellation of surfaces (such as tiling a plane like your bathroom floor). Instead, our intent is to capture the notion of *motif* as it is used in biological pattern recognition, most commonly in DNA and protein sequence motives.

Officially, *motif* may be pluralized either *motives* (with the accent on the last syllable) or *motifs*. One will find both in the biological literature.

See also *biochemical, chemical, dynamical, functional, kinetic, mechanistic, phylogenetic, regulatory, thermodynamic*, and *topological motives*.

N

Navier-Stokes equations Let $U \subset \mathbb{R}^n$ be open and $u : U \times \mathbb{R} \to \mathbb{R}^n$. The *Navier-Stokes equations* for incompressible, viscous flows are

$$\begin{cases} u_t + u \cdot Du = -Dp \\ div\, u = 0 \end{cases}$$

where $Du = D_x u = (u_{x_1}, \ldots, u_{x_n})$ denotes the gradient of u with respect to the spatial variable $x = (x_1, \ldots, x_n)$.

neighborhood (of a point p in a *topological space*) A set containing an open set containing p.

nematic phase See *liquid-crystal transitions*.

network A mathematical graph $N(\mathcal{V}, \mathcal{E}, \mathcal{P}, \mathcal{L})$ consisting of a set of *nodes* (or vertices) $\mathcal{V}$, a set of *edges* between the *nodes* $\mathcal{E}$, a set of parameters $\mathcal{P}$ describing properties of subgraphs of the network, and a set of labels $\mathcal{L}$.

Comment: If $\mathcal{P} = \emptyset$ the network reduces to its corresponding graph [denoted either $N'(\mathcal{V}, \mathcal{E}, \mathcal{L})$ or $G(\mathcal{V}, \mathcal{E}, \mathcal{L})$ according to one's taste]. In general, all graph algorithms apply to networks, but there are some specialized algorithms, relying on the parameters, which do not apply to graphs.

nodal basis The basis $\{b_1, \cdots, b_N\}$ of a *finite element space* V_h dual to the set Π_h of global *degrees of freedom*, that is, $\phi_k(b_i) = \delta_{ik}, i, k = 1, \cdots, N$, is called the *nodal basis* of V_h. Thanks to the localization of the *degrees of freedom*, the *nodal basis* functions are locally supported. More precisely, supp b_i is contained in the closure of the union of all those cells of Ω_h that share the *vertex, face, edge*, etc. to which the related global *degree of freedom* is associated.

nodal interpolation operator Given a *finite element space* V_h with set Π_h of nodal *degrees of freedom*, we can define a projection

$$I_h : S \cup V_h \to V_h, \; \phi(I_h u) = \phi(u) \; \forall \phi \in \Pi_h,$$

where S is a space of sufficiently *smooth* functions/vector fields for which the evaluation of *degrees of freedom* is well defined. Note that for many finite element spaces that are intended as conforming approximating spaces for a function space V, S will be strictly contained in V. Nodal interpolation operators are local in the sense that for each cell K of the underlying *triangulation* the restriction of $I_h u$ onto K depends on $u_{|\bar{K}}$.

node A *node*, v_i (i indexing the nodes of the graph), is a member of the set of nodes ($\mathcal{V}$) of a mathematical graph or network.

Comment: Each node is commonly rendered by a geometric point, but in fact it is simply an element of any set; so in a set of integers, each integer would be a node of a graph. A common synonym is *vertex*. When reactions and compounds are both nodes in the representation of a biochemical network, because there are two types of nodes, the network is said to be *bipartite*.

nonconforming finite elements A *finite element space* V_h is called nonconforming with respect to a function space V if $V_h \not\subset V$. Using V_h to discretize a variational problem over V amounts to a *variational crime*. However, it might be feasible, if V_h satisfies certain consistency conditions expressed in *Strang's second lemma*.

nondeterministic computation Let the sets of presented inputs and produced outputs be $\Sigma_{\mathcal{I}'}$ and $\Sigma_{\mathcal{O}'}$, respectively; the cardinalities of sets be denoted by the double overset bars; and $\sigma_{i,\mathcal{I}'}$ and $\sigma_{i,\mathcal{O}'}$ be the particular symbols presented to or produced by the mapping c_i.

Then a *nondeterministic computation* $\mathcal{C}_n$ specifies a computation $\mathcal{C}: \Sigma_{\mathcal{I}'} \to \Sigma_{\mathcal{O}'}$, such that $\overline{\overline{\Sigma}}_{\mathcal{I}'} \geq 1; \overline{\overline{\Sigma}}_{\mathcal{O}'} \geq 1$; at least one mapping $c_i \in \mathcal{C}_n$ transforming presented inputs to presented outputs is many-to-many; and $\forall \sigma_{i,\mathcal{I}'}, \sigma_{i,\mathcal{I}'} \in \Sigma_{\mathcal{I}'}, \forall \sigma_{i,\mathcal{O}'}, \sigma_{i,\mathcal{O}'} \in \Sigma_{\mathcal{O}'}$, the probabilities that each exists, $P_e(\sigma_{i,\mathcal{I}'})$ and $P_e(\sigma_{i,\mathcal{O}'})$, are unity.

Comment: This computation is not stochastic, but it is nondeterministic. There is at least one step in the computation where the output produced is not a function of the input presented, even though all inputs and outputs exist with probability one. See *deterministic* and *stochastic computations*.

nonenzymatic degree   The degree of a reactive conjunction node excluding the *macromolecular* catalyst.

nonlinear optical effect   An effect brought about by electromagnetic radiation the magnitude of which is not proportional to the irradiance. Nonlinear optical effects of importance to photochemistry are *harmonic frequency generation*, lasers, Raman shifting, *upconversion*, and others.

norm (of a vector)   See *angle between vectors*, *normed space*.

normal   Perpendicular, as a vector to a surface or to another vector.

normed space   A vector space X with a *norm* $||x||$ defined, for $x \in X$, satisfying

  (i.) $||x|| \geq 0$ and equality only when $x = 0$,

  (ii.) $||x + y|| \leq ||x|| + ||y||$, and

  (iii.) $||\lambda x|| = |\lambda| \, ||x||$.

Nöther's theorem   (1) If the *Lie algebra* **g** acts canonically on the *Poisson manifold* P, admits a *momentum map* $J : P \to \mathbf{g}^*$, and $H \in C^\infty(P)$ is **g**-invariant, i.e., $\xi_P(H) = 0$ for all $\xi \in \mathbf{g}$, then J is a constant of the motion for H, i.e.,

$$J \circ \varphi_t = J$$

where φ_t is the *flow* of the *Hamiltonian vector field* X_H.

(2) (of a *Lagrangian system*) A map between 1-parameter families of *Lagrangian symmetries* and *conserved currents*. For example, let (M, L) be a *time-independent Lagrangian system* and $X = X^\lambda \partial_\lambda$ an infinitesimal generator of

symmetries; the associated *conserved current* is given by

$$\mathcal{E} = \frac{\partial L}{\partial u^\lambda} X^\lambda$$

where (x^λ, u^λ) are local coordinates on the tangent bundle TM.

nuclear decay   A spontaneous nuclear transformation.

nuclear fission   The division of a nucleus into two or more parts with masses of equal order of magnitude, usually accompanied by the emission of neutrons, gamma radiation, and, rarely, small charged nuclear fragments.

nucleic acids   *Macromolecules*, the major organic matter of the nuclei of biological cells, made up of *nucleotide* units, and hydrolyzable into certain *pyrimidine* or *purine bases* (usually adenine, cytosine, guanine, thymine, or uracil), D-ribose or 2-deoxy-D-ribose, and phosphoric acid.
    See *deoxyribonucleic acids*.

nucleosides   Ribosyl or deoxyribosyl derivatives (rarely, other glycosyl derivatives) of certain *pyrimidine* or *purine bases*. They are thus *glycosylamines* or *N-glycosides* related to *nucleotides* by the lack of phosphorylation. It has also become customary to include among *nucleosides* analogous substances in which the *glycosyl group* is attached to carbon rather than nitrogen ("C-nucleosides"). See also *nucleic acids*.

nucleotides   Compounds formally obtained by esterification of the 3' or 5' hydroxy group of *nucleosides* with phosphoric acid. They are the *monomers* of *nucleic acids* and are formed from them by hydrolytic cleavage.

nuclide   A species of atom, characterized by its mass number, atomic number, and nuclear energy state, provided that the mean life in that state is long enough to be observable.

O

oligomer   A substance composed of *oligomer molecules*. An *oligomer* obtained by telomerization is often termed a telomer.

oligomer molecule   A molecule of intermediate relative molecular mass, the structure of which essentially comprises a small plurality of units derived, actually or conceptually, from molecules of lower relative molecular mass.

Notes: (**1**) A molecule is regarded as having an intermediate relative molecular mass if it has properties which do vary significantly with the removal of one or a few of the units.

(**2**) If a part or the whole of the molecule has an intermediate relative molecular mass and essentially comprises a small plurality of units derived, actually or conceptually, from molecules of lower relative molecular mass, it may be described as oligomeric, or by oligomer used adjectivally.

one-dimensional Ising model   One of the most important models in statistical mechanics. It consists of a set of identical spins $\{..., S_{-2}, S_{-1}, S_0, S_1, S_2, ...\}$ with nearest neighbor interaction energy $J S_i S_{i+1}$. Each spin can take values ± 1. An infinite *one-dimensional Ising system* can exhibit phase transition behavior.

onto   A mapping or transformation of a set X which transforms points of X to those in Y is a mapping of X *onto* Y if each point of Y is the image of at least one point of X. For example, $y = 3x + 2$ is a mapping of the real numbers onto the real numbers; $y = x^2$ is a mapping of the real numbers into the real numbers, or onto the non-negative real numbers. See also *into, bijection,* and *injection.*

open covering (of a topological space $[X, \tau(X)]$) A family $\{U_\alpha\}_{\alpha \in I}$ of open sets in X such that $\bigcup_{\alpha \in I} U_\alpha = X$ where I is any set of indices. On a *paracompact manifold M* every open covering contains a *locally finite covering*, i.e., an open

covering $\{U_\alpha\}_{\alpha \in I}$ such that for any $x \in M$ there is an open *neighborhood* U_x which intersects just a finite number of elements U_α of the covering. On a *paracompact manifold M* every open covering contains a *good covering*, i.e., an open covering $\{U_\alpha\}_{\alpha \in I}$ such that for any finite set of indices $\alpha_1, \alpha_2, \ldots, \alpha_k \in I$ the intersection $U_{\alpha_1 \alpha_2 \ldots \alpha_k} = \bigcap_{i=0}^{k} U_{\alpha_i}$ (which is an open set by definition) is *topologically trivial*, i.e., it is *contractible*.

open neighborhood (of a point x of a *topological space* $[X, \tau(X)]$)   An *open subset* $U \subset X$ containing x.

open set   A subset $U \subset X$ of a *topological space* X which is an element of its *topology* $\tau(X)$.

In $\mathbb{R}^n$, with the standard metric topology, a subset $U \subset \mathbb{R}^n$ is open if any point $x \in U$ is contained in U together with an open n-ball centered at x of radius r, i.e., $B_x^r = \{y \in \mathbb{R}^n : |y-x| < r\}$.

orbit of an action   See *left action* and *right action*.

order of reaction, *n*   If the macroscopic (observed, empirical, or phenomenological) *rate of reaction* (v) for any reaction can be expressed by an empirical differential rate equation (or rate law) which contains a factor of the form $k[A]^\alpha[B]^\beta$ ... (expressing in full the dependence of the rate of reaction on the concentrations [A],[B]...) where α, β are constant exponents (independent of concentration and time) and k is independent of [A] and [B], etc. (rate constant, rate coefficient), then the reaction is said to be of order α with respect to A, of order β with respect to B,..., and of (total or overall) order $n = \alpha + \beta + \ldots$. The exponents $\alpha, \beta, \ldots$ can be positive or negative integral or rational nonintegral numbers. They are the reaction orders with respect to A,B, ... and are sometimes called "partial orders of reaction." Orders of reaction deduced from the dependence of initial rates of reaction on concentration are called "orders of reaction with respect to concentration"; orders of reaction deduced from the dependence of the rate of reaction on time are called "orders of reaction with respect to time."

The concept of order of reaction is also applicable to chemical rate processes occurring in systems for which concentration changes (and hence the rate of reaction) are not themselves measurable, provided it is possible to measure a *chemical flux*. For example, if there is a dynamic equilibrium according to the equation:

$$aA \rightleftharpoons pP$$

and if a chemical flux is experimentally found, (e.g., by NMR *line-shape analysis*) to be related to concentrations by the equation

$$\phi - A/\alpha = k[A]^{\alpha}[L]^{\lambda}$$

then the corresponding reaction is of order α with respect to A,... and of total (or overall) order $n(= \alpha + \lambda + \ldots)$. The proportionality factor k above is called the (nth order) "rate coefficient." Rate coefficients referring to (or believed to refer to) *elementary reactions* are called "rate constants" or, more appropriately, "microscopic" (hypothetical, mechanistic) rate constants.

The (overall) order of a reaction cannot be deduced from measurements of a "rate of appearance" or "rate of disappearance" at a single value of the concentration of a species whose concentration is constant (or effectively constant) during the course of the reaction. If the overall rate of reaction is, for example, given by

$$\nu = k[A]^{\alpha}[B]^{\beta}$$

but [B] stays constant, then the order of the reaction (with respect to time), as observed from the concentration change of A with time, will be α, and the rate of disappearance of A can be expressed in the form

$$\nu_A = k_{obs}[A]^{\alpha}.$$

The proportionality factor k_{obs} deduced from such an experiment is called the "observed rate coefficient," and it is related to the $(\alpha + \beta)$th order rate coefficient k by the equation:

$$k_{obs} = k[B]^{\beta}.$$

For the common case when $\alpha = 1$, k_{obs} is often referred to as a "pseudo-first-order rate coefficient" (k_{ψ}).

For a simple (*elementary*) reaction a partial order of reaction is the same as the *stoichiometric* number of the reactant concerned and must therefore be a positive integer. The overall order is then the same as the *molecularity*. For *stepwise reactions* there is no general connection between stoichiometric numbers and partial orders. Such reactions may have more complex rate laws, so that an apparent *order of reaction* may vary with the concentrations of the *chemical species* involved and with the progress of the reaction. In such cases, it is not useful to speak of *orders of reaction*, although apparent orders of reaction may be deducible from initial rates. In a *stepwise reaction*, *orders of reaction* may in principle always be assigned to the elementary steps.

ordering (of a set) **(1)** Preordering: a relation $\triangleright$ on a set A such that

  (i.) $\forall a \in A \; a \triangleright a$;

  (ii.) $\forall a, b, c \in A \; a \triangleright b, b \triangleright c \Rightarrow a \triangleright c$.

(2) Partial ordering: a preordering such that

  (i.) $\forall a, b \in A \; a \triangleright b$ and $b \triangleright a \Rightarrow a = b$.

(3) Total ordering: a partial ordering such that

  (i.) $\forall a, b \in A \; a \triangleright b$ or $b \triangleright a$.

Examples: The inclusion is a partial ordering in the *power set* $\mathcal{P}(X)$ of a set X. The relation $\geq$ is a total ordering in the real line $\mathbb{R}$. The relation $z \triangleright w$ if and only if $|z| \geq |w|$ defined on the complex plane $\mathbb{C}$ is a preordering but not a partial ordering.

oregonator R.M. Noyes and R.J. Field at the University of Oregon developed a mathematical model, consisting of three coupled nonlinear ordinary differential equations, for the BZ reaction (see *Belousov-Zhabotinskii reaction*). The model was shown to have a limit cycle, hence firmly established the theoretical basis of chemical oscillation (*cf.* R.M. Noyse, *J. Chem. Educ.*, 66, 190, 1989).

orientation (of an m-dimensional *manifold M*) A nondegenerate m-form over M. If (M, g) is a (pseudo)-*Riemannian manifold* and $\mathrm{ds} = \mathrm{d}x^1 \wedge \mathrm{d}x^2 \wedge \cdots \wedge \mathrm{d}x^m$ is the canonical local basis of m-forms over M, then $\eta = \sqrt{g}\mathrm{ds}$ is an orientation.

The *manifolds* that allow orientations are called *orientable*, and they allow *oriented atlases*, i.e., atlases with *transition functions* with definite positive Jacobians.

Ornstein-Uhlenbeck process   A Gaussian, Markov stochastic processes defined by a linear stochastic differential equation

$$dX = -bXdt + adW(t)$$

in which $a, b > 0$, and $W(t)$ is the Wiener process, i.e., dW/dt is the white noise.

orthogonal   Describing two elements V_1, V_2 of a *Hilbert space* such that $\langle v_1, v_2 \rangle = 0$.

orthogonal group   The group $O(m)$ is the group of isometries of $\mathbb{R}^m$ with the standard positive definite metric δ. Let us choose an orthonormal basis E_i in $\mathbb{R}^m$, the metric δ is expressed by $\delta = \delta_{ij} E^i \otimes E^j$ (where E^i is the *dual basis*). An *endomorphism* $\alpha : \mathbb{R}^m \to \mathbb{R}^m$ is an element of the *orthogonal group* $O(m)$ if and only if

$$A_i^a \, \delta_{ab} \, A_j^b = \delta_{ij} \qquad \alpha(E_i) = A_i^j \, E_j.$$

Thence the inverse of an orthogonal matrix coincides with its transpose. Example: Rotations in $\mathbb{R}^3$ (as well as reflections) are elements of $O(3)$.

Analogously, the orthogonal group $O(r, s)$ (with $r + s = m$) is the group of isometries of the indefinite metric of signature (r, s) on $\mathbb{R}^m$. Example: The *Lorentz group* is the *orthogonal group* $O(1, m - 1)$.

The subgroup of isometries $\alpha : \mathbb{R}^m \to \mathbb{R}^m$ which preserve the *orientation* of $(\mathbb{R}^m, \eta)$ is denoted by $SO(r, s)$, and it is called the *special orthogonal group*.

orthonormal basis   An orthonormal set which is also a basis. Synonym: complete orthonormal set.

orthonormal set   A set $\{u_\alpha\}$ of a *Hilbert space H* satisfying

$$\langle u_\alpha, u_\beta \rangle = \begin{cases} 0 & \text{if } \alpha \neq \beta \\ 1 & \text{if } \alpha = \beta \end{cases}.$$

P

p-Laplacian equation   Let $U \subset \mathbb{R}^n$ be open and $u : U \times \mathbb{R} \to \mathbb{R}$. The *p-Laplacian equation* for u is

$$div(|Du|^{p-2} Du) = 0,$$

where $Du = D_x u = (u_{x_1}, \ldots, u_{x_n})$ denotes the gradient of u with respect to the spatial variable $x = (x_1, \ldots, x_n)$.

p-version of finite elements   The finite element discretization of a boundary value problem is built upon a fixed *mesh*. A family of finite elements is employed that supplies *finite element spaces* for a wide range of local polynomial degrees. The idea of the p-version of finite elements is to get an acceptable discrete solution by using a sufficiently high degree polynomial. A local variant of the p-version raises the polynomial degree only on cells where a better approximation is really needed to reduce the discretization error; *cf.*, *adaptive refinement*.

parallel transport   A vector field X defined along a *curve* γ in a *manifold* M with a connection Γ such that

$$\nabla_{\dot{\gamma}} X = 0$$

where $\dot{\gamma}$ denotes the *tangent* vector to the curve γ.

parametric equivalence   Two *finite elements* (K, V_K, Π_K) and (T, V_T, Π_T) are *parametric equivalent*, if there is a piecewise *smooth bijective mapping* $\Phi : K \to T$ and a *bijective linear mapping* $\mathcal{F}_\Phi : V_T \to V_k$, the *pull-back*, such that

$$\Pi_T = \{ \kappa : V_T \to \mathbb{C}, \kappa(u) = \phi(\mathcal{F}_\Phi u), \phi \in \Pi_k \}.$$

Sloppily speaking, this means that both the geometric elements and the local spaces can be mapped onto each other, and that the local *degrees of freedom* are invariant with respect

to pull-back. An important consequence is that *nodal interpolation operators* and pull-backs commute:

$$\mathcal{F}_\Phi(I_h u_{|T}) = I_h(\mathcal{F}_\Phi(u))_{|K},$$

for all sufficient smooth functions/vector fields u on T.

Given an infinite family of *meshes* $\{\Omega_h\}_{h \in \mathbb{H}}$, $\mathbb{H}$ an index set, a related family of *finite element spaces* is considered parametric equivalent, if there is a small number of finite elements to which all elements of the family are *parametric equivalent*. These elements are called *reference elements*.

Parametric equivalence is a key tool in most proofs of local a priori interpolation estimates for families of finite element spaces and their associated *nodal interpolation operators*. If the *meshes* are *shape regular*, it is often possible to gauge the change of norms under pull-back. First the interpolation error is estimated on the reference element(s). Then, the relationship of commutativity given above is used to conclude an estimate for an arbitrary element.

partial differential equation   A *differential equation* involving a function of more than one variable, and hence partial derivatives.

partition   A set of subgraphs, $\{ \mathsf{G}'(\mathcal{V}', \mathcal{E}'),$ $\mathsf{G}''(\mathcal{V}'', \mathcal{E}''), \ldots \}$ of a graph $\mathsf{G}(\mathcal{V}, \mathcal{E})$, such that the subgraphs are disjoint.

partition of unity   Let M be a paracompact *manifold* and $\{U_\alpha\}_{\alpha \in I}$ a *locally finite covering* (see *open covering*). A *partition of unity relative to* $\{U_\alpha\}_{\alpha \in I}$ is a family of (*smooth*) local functions $f_\alpha : M \to \mathbb{R}$ such that:

(i.) $\forall x \in M, f_\alpha(x) \geq 0$;

(ii.) $\operatorname{supp} f_\alpha \subset U_\alpha$;

(iii.) $\forall x \in M, \sum_{\alpha \in I} f_\alpha(x) = 1$;

where $\operatorname{supp} f_\alpha$ denotes the *support* of the function f_α, i.e., the closure of the set $\{ x \in M : f_\alpha(x) \neq 0 \}$.

Notice that the sum in *P3* is finite since the covering $\{U_\alpha\}_{\alpha \in I}$ is locally finite. The functions f_α of a partition of unity can be required to be real-analytic only in trivial situations. In fact, any analytical function which is identically zero outside U_α is zero everywhere. On the contrary, the functions f_α can be required to satisfy less stringent regularity conditions such as C^∞.

Partitions of unity are often used to prove existence of a global object once local objects are known to exist in M. For example, let M be a (paracompact) C^∞ *manifold* and $\{(U_\alpha, x_\alpha)\}_{\alpha \in I}$ an atlas of M. This means that U_α are *diffeomorphic* (via the maps $x_\alpha : U_\alpha \to \mathbb{R}^m$) to an open set $x_\alpha(U_\alpha) \subset \mathbb{R}^m$ ($m = \dim(M)$). Consequently, there exist local (strictly) *Riemannian* metrics g_α induced on U_α (via the maps $x_\alpha : U_\alpha \to \mathbb{R}^m$) by the standard metric $\delta_{\alpha\beta}$ on $\mathbb{R}^m$. Notice that metrics are second rank, nondegenerate, positive definite tensors. If $\{(U_\alpha, f_\alpha)\}$ is a partition of unity relative to the open covering $\{U_\alpha\}$, we can then define the second-order tensors $f_\alpha \, g_\alpha$, one for each $\alpha \in I$. They are global tensors but they identically vanish outside U_α so that they are not suitable to define a *Riemannian* metric over M. Let us, however, consider the combination $g = \sum_{\alpha \in I} f_\alpha \, g_\alpha$ (which exists since $\{U_\alpha\}$ is locally finite). Then g is a global *Riemannian* metric over M, since linear combinations with positive coefficients of positive definite tensors are still positive definite. The same argument cannot be applied in general to arbitrary signature and in particular to the Lorentzian case. In those cases, in fact, further topological conditions have to be satisfied for a global metric to exist.

parts of collections   For any bag, list, sequence, or set, a *subpart* (subbag, sublist, subsequence, or subset) is a portion of the original collection. If the part is less than the original collection, it is a *proper subbag, sublist, subsequence,* or *subset*, and we denote the relationship between the part and the collection by part $\subset$ collection. Otherwise, the relationship is denoted by $\subseteq$ to indicate the part may be equal to the whole.

Comment: A set, bag, list, or sequence may contain another of its type. For example, $\{a, b\} \subset \{a, b, c\}$ and $\langle a, b \rangle \subset \langle a, b, c \rangle$.

In forming parts of collections, bear in mind that the part must satisfy the same properties as the whole. For example, if we have $S = \langle a, b, c, d, e \rangle$, then any derived subsequence must have the same precedence relations: $\{\langle a, b \rangle, \langle b, c, d \rangle, \langle d, e \rangle\}$ is a set of

valid subsequences, but $\langle d, c \rangle$ is not a subsequence of S. See also *bag, empty collection, list, sequence, set,* and *superset*.

path   An alternating sequence of *nodes* and *edges* drawn from a graph $\mathsf{G}(\mathcal{V}, \mathcal{E})$, such that they form a connected subgraph, $\mathsf{G}'(\mathcal{V}', \mathcal{E}')$, where $\mathcal{V}' \subseteq \mathcal{V}$ and $\mathcal{E}' \subseteq \mathcal{E}$.

Comment: A path through a graph is simply a connected subgraph whose *nodes* and *edges* appear in sequence: just begin at the beginning and walk along the path to the end. If one permits oneself to let the *nodes* of the path be implicitly represented by the *edges*, the path becomes a sequence of *edges*. See also *pathway, sequence,* and *terminal nodes*.

pathway   A sequence of biochemical reactions and their compounds whose *nodes* and *edges* form a path and have historically been considered by biochemists to be a *pathway*.

Comment: This definition seems a little circular, but in fact what we define as biochemical pathways is largely determined by the history and results of the experiments involved in their discovery. Definitions of particular pathways vary slightly among different sources. For example, some authors include phosphorylation of D-glucopyranose as part of glycolysis; others do not. Still others refer to the next step as the first "committed step" in glycolysis. See also *path, sequence,* and *terminal nodes*.

pathwise connected   The property of a topological space $(X, \tau(X))$ that any two points can be joined by a *curve*. See *connected*. Pathwise connectedness implies connectedness. The converse is false.

Example: The subset in $\mathbb{R}^2$ given by the union of the set $L = \{(0, x) : x \in [0, 1]\}$ and the set $S = \{(x, \sin(\frac{1}{x})) : 0 \leq x\}$ is connected but not *pathwise connected*.

pattern   See *motif*.

pendant node   See *singleton node*.

pericyclic reaction A *chemical reaction* in which *concerted* reorganization of bonding takes place throughout a cyclic array of continuously bonded atoms. It may be viewed as a reaction proceeding through a fully *conjugated* cyclic transition state. The number of atoms in the cyclic array is usually six, but other numbers are also possible. The term embraces a variety of processes, including *cycloadditions, cheletropic reactions, electrocyclic reactions*, and *sigmatropic rearrangements*, etc. (provided they are concerted).

photoelectrical effect The ejection of an electron from a solid or a liquid by a photon.

photoelectron spectroscopy (PES) A spectroscopic technique that measures the kinetic energy of electrons emitted upon the ionization of a substance by high energy monochromic photons. A photoelectron spectrum is a plot of the number of electrons emitted vs. their kinetic energy. The spectrum consists of bands due to transitions from the ground state of an atom or molecular entity to the ground and excited states of the corresponding radical cation. Approximate interpretations are usually based on "Koopmans theorem" and yield orbital energies. PES and UPS (UV photoelectron spectroscopy) refer to the spectroscopy using vacuum ultraviolet sources, while ESCA (electron spectroscopy for chemical analysis) and XPS use X-ray sources.

photoemissive detector A *detector* in which a *photon* interacts with a solid surface, which is called the photocathode, or a gas, releasing a photoelectron. This process is called the *external photoelectric effect*. The photoelectrons are collected by an electrode at positive electric potential, i.e., the anode.

photon Particle of zero charge, zero rest mass, spin quantum number 1, energy $h\nu$, and momentum $h\nu/c$ (h is the Planck constant, ν the frequency of radiation, and c the speed of light); carrier of electromagnetic force.

photon flow, Φ_p The number of *photons* (quanta, N) per unit time. (dN/dt, simplified expression: $\Phi_p = N/t$ when the number of photons is constant over the time considered.) The SI unit is s^{-1}. Alternatively, the term can be used with the amount of *photons* (mol or its equivalent Einstein), the SI unit then being mol s^{-1}.

photon fluence, H_p^0 The integral of the amount of all *photons* (quanta) which traverse a small, transparent, imaginary spherical target, divided by the cross-sectional area of this target. The *photon* fluence rate, E_p^0, integrated over the duration of the irradiation ($\int E_p^0 dt$, simplified expression: $H_p^0 = E_p^0 t$ when E_p^0 is constant over the time considered). *Photons* per unit area (quanta m^{-2}). The SI unit is m^{-2}. Alternatively, the term can be used with the amount of *photons* (mol or its equivalent Einstein), the SI unit then being mol m^{-2}.

photon fluence rate, E_p^0 The rate of *photon fluence*. Four times the ratio of the *photon flow*, Φ_p, incident on a small, transparent, imaginary spherical volume element containing the point under consideration divided by the surface of that sphere, S_K. ($\int_{4\pi} L_p dw$, simplified expression: $E_p^0 = 4\Phi_p/S_K$ when the *photon flow* is constant over the solid angle considered). The SI unit is $m^{-2}s^{-1}$. Alternatively, the term can be used with the amount of *photons* (mol or its equivalent Einstein), the SI unit then being mol $m^{-2}\ s^{-1}$. It reduces to *photon irradiance* for a parallel and normally incident beam not scattered or reflected by the target or its surroundings.

photon flux Synonymous with *photon irradiance*.

photon irradiance, E_p The *photon flow*, Φ_p, incident on an infinitesimal element of surface containing the point under consideration divided by the area (S) of that element ($d\Phi_p/ds$, simplified expression: $E_p = \Phi_p/S$ when the *photon flow* is constant over the surface considered). The SI unit is $m^{-2}s^{-1}$. Alternatively, the term can be used with the amount of *photons* (mol or its equivalent Einstein), the SI unit then being $mol^{-2}s^{-1}$. For a parallel and perpendicularly incident beam not scattered or reflected by the target or its surroundings *photon fluence rate* (E_p^0) is an equivalent term.

phylogenetic motif Any *motif* conserved over biological species. See also *biochemical, chemical, dynamical, functional, kinetic, mechanistic, regulatory, thermodynamic,* and *topological motives.*

physical quantity (measurable quantity) An attribute of a phenomenon, body, or substance that may be distinguished qualitatively and determined quantitatively.

planar graph A graph that can be drawn in the plane with no *edges* crossing.

plasmid An extrachromosomal genetic element consisting generally of a circular duplex of *DNA* which can replicate independently of chromosomal DNA. R-plasmids are responsible for the mutual transfer of antibiotic resistance among microbes. *Plasmids* are used as *vectors* for cloning DNA in bacteria or yeast host cells.

ploidy A term indicating the number of sets of *chromosomes* present in an organism, e.g., haploid (one) or diploid (two).

Poisson-Boltzmann equation A mathematical model for the ionic gas (plasma) or ionic solution. The nonlinear equation is established based on two physical laws: the Poisson equation relating the electric potential to charge distribution, and the Boltzmann law relating the charge distribution to electric potential. It has been shown that this equation is a good model for many applications even though it suffers from some *thermodynamic* inconsistency. The linearized equation and its solution is known as Debye-Hückel theory. The one-dimensional version of the nonlinear equation can be applied to modeling the charge distribution near a flat, charged membrane; this is known as Guy-Chapman equation.

Poisson bracket Let (P, ω) be a *symplectic manifold*, ξ^A a system of canonical coordinates and f, g two real functions on P. Then the *Poisson bracket* of f and g is defined as the function

$$\{f, g\} = \omega^{-1}(\mathrm{d}f, \mathrm{d}g) = \omega^{AB}(\partial_A f)(\partial_B g).$$

Functions on P with *Poisson bracket* form a *Lie algebra*. There exists a *Lie algebra homomorphism* between the *Lie algebra* of functions on P and vector fields, given by

$$H \mapsto X_H(f) = \{H, f\}.$$

It is an *isomorphism* when constants are quotiented out of functions.

More generally, let P be a *smooth manifold* and $C^\infty(P)$ the space of smooth functions. A *Poisson bracket* or *Poisson structure* on P is an operation $\{\ ,\ \} : C^\infty(P) \times C^\infty(P) \to C^\infty(P)$ satisfying the following

(i.) $\{F, G\}$ is real, bilinear in F and G,

(ii.) $\{F, G, \} = -\{G, F\}$, skew symmetric,

(iii.) $\{\{F, G\}, H\} + \{\{H, F\}, G\} + \{\{G, H\}, F\} = 0$, Jacobi identity,

(iv.) $\{FG, H\} = F\{G, H\} + \{F, H\}G$, Leibniz rule.

Poisson equation Let $U \subset \mathbb{R}^n$ open and $u : U \times \mathbb{R} \to \mathbb{R}$. The (nonlinear) *Poisson equation* for u is

$$-\Delta u = f(x).$$

Poisson manifold A *smooth manifold* P with a *Poisson bracket* $\{\ ,\ \}$. Examples: *Symplectic manifolds* are *Poisson manifolds*, where the *Poisson bracket* is defined by

$$\{F, G\}(x) = \omega(x)(X_F(x), X_G(x)), \ x \in M$$

where X_F is the *Hamiltonian vector field* of F.

polar coordinates The parameterization of the plane $\mathbb{R}^2$ as $X = (r, \theta)$, where $r > 0$ is the absolute value $|X|$ and θ is the angle (in radians) between the horizontal axis and the segment from O to X.

pole A complex number z_0 is a *pole* of a function $f(z)$ if $f(z)$ is analytic in $0 < |z - z_0| < \epsilon$, for some $\epsilon > 0$ and not analytic at z_0, but $(z - z_0)^n f(z)$ is analytic at z_0, for some positive integer n.

pollution effect   If some feature of a problem has an indirect and nonobvious adverse impact on the accuracy that can be achieved by a discretization scheme, a *pollution effect* can be observed. A prominent example is the deterioration of certain asymptotic rates of convergence of finite element schemes for second-order elliptic boundary value problems in the presence of reentrant corner. Another case is the loss of accuracy suffered by standard finite element schemes for the Helmholtz equation $-\Delta u - \kappa u = f$ when κ becomes large.

polymer   A substance composed of *macromolecules*.

polymerization   The process of converting a *monomer* or a mixture of monomers into a *polymer*.

polynomial   A linear combination of nonnegative, integer powers of a variable (or unknown) x,

$$p(x) = a_0 + a_1 x + \cdots + a_n x^n.$$

population genetics   One can study genetics from either a molecular point of view (see *double helix*) or from a population point of view. Population genetics studies where the gene is located in a chromosome, and how it is passed from generation to generation from the pedigree of hereditary markers.

population model   A system of mathematical equations representing the proposed relation between the population sizes and their growth rates. The often used equations in textbooks, such as exponential growth, logistic equation, predator-prey, and competition are simplified population models.

porous medium equation   Let $U \subset \mathbb{R}^n$ open and $u : U \times \mathbb{R} \to \mathbb{R}$. The *porous medium equation* for u is

$$u_t - \Delta(u^\gamma) = 0.$$

Post production system (PPS)   A finite grammar Π over a finite set of symbols Σ which produces a set of constructs, $\mathcal{C}$, from an initial construct $c_0 \in \mathcal{C}$, such that each construct in $\mathcal{C}$ can be obtained from one or more symbols in Σ by a finite derivation, where each step in the derivation is a rule $\pi_i \in \Pi$.

Comment: The constructs were strings representing logical predicates, and the grammar the rules of logical derivation and proof, in Post's original formulation (E.L. Post, "Finite combinatory processes. Formulation I", *J. Symbolic Logic*, **1**, 103–105 (1936)). The class of Post-generated construct sets is a member of the class of recursively enumerable sets. The grammar may be deterministic or nondeterministic, but its execution is not stochastic. See *universal Turing machine* and *von Neumann machine*.

postfix   An operator written after its operands. Thus for two operands x and y and operator f, the syntax is $x\,y\,f$.

Comment: Remember calculators with reverse Polish notation? In many instances this is the most natural order of all; for example, exponentiation. See also *functor, infix, prefix*, and *relation*.

posynomial   A positive sum of polynomials $\sum_i c_j \Pi_j x(j)^{a(i,j)}$, where $c_j > 0$. Each monomial is the product

$$\Pi_j x(j)^{a(i,j)} = x(1)^{a(i,1)} \cdot x(2)^{a(i,2)} \cdot \ldots \cdot x(n)^{a(i,n)}$$

and $[a(i, j)]$ is called the exponent matrix. This is the fundamental function in geometric programming.

potential energy, E_p, V   Energy of position or orientation in a field of force.

potential-energy profile   A *curve* describing the variation of the potential energy of the system of atoms that make up the reactants and products of a reaction as a function of one geometric coordinate, and corrresponding to the "energetically easiest passage" from reactants to products (i.e., along the line produced by joining the paths of steepest descent from the *transition state* to the reactants and to the products). For an *elementary reaction*, the relevant geometric coordinate is the *reaction coordinate*. For a *stepwise reaction* it is the succession of reaction coordinates for the successive individual reaction steps. (The reaction coordinate is sometimes approximated by a quasi-chemical index of reaction progress, such as degree of atom transfer or *bond order* of some specified bond.)

potential-energy (reaction) surface   A geometric hypersurface on which the potential energy of a set of reactants is plotted as a function of the coordinates representing the molecular geometries of the system.

For simple systems two such coordinates (characterizing two variables that change during the progress from reactants to products) can be selected, and the potential energy plotted as a contour map.

For simple *elementary reactions*, e.g., $A - B + C \rightarrow A + B - C$, the surface can show the potential energy for all values of the A,B,C geometry, providing that the ABC angle is fixed.

For more complicated reactions a different choice of two coordinates is sometimes preferred, e.g., the *bond orders* of two different bonds. Such a diagram is often arranged so that reactants are located at the bottom left corner and products at the top right. If the trace of the representative point characterizing the route from reactants to products follows two adjacent edges of the diagram, the changes represented by the two coordinates take place in distinct succession. If the trace leaves the edges and crosses the interior of the diagram, the two changes are *concerted*. In many qualitative applications it is convenient (although not strictly equivalent) for the third coordinate to represent the standard Gibbs energy rather than potential energy.

Using bond orders is, however, an oversimplification, since these are not well defined, even for the transition state. (Some reservations concerning the diagrammatic use of Gibbs energies are noted under *Gibbs energy diagram*.)

The energetically easiest route from reactants to products on the potential-energy contour map defines the *potential-energy profile*.

power series   A formal summation of the form $\sum_{n=0}^{\infty} c_n (x - a)^n$, where a and $c_0, c_1, \dots$ are complex numbers.

power set (of a set A)   The set of all subsets of A (the empty subset included). It is denoted by $\mathcal{P}(A)$ or 2^A. For example, the empty set $\emptyset$ has no element. Its power set $\mathcal{P}(\emptyset) = \{\emptyset\}$ has exactly one element. The power set of $A = \{a, b\}$ is $\mathcal{P}(A) = \{\emptyset, \{a\}, \{b\}, A\}$. In general, if A has n elements, the power set has 2^n elements.

predator-prey population model   A mathematical model, in terms of a pair of ordinary differential equations, representing the populations of two species with prey-predator relation. Such systems are likely to exhibit oscillation or even limit cycles. Generalization to more than two species is possible and, in fact, is widely used in applications.

prefix   An operator written before its operands. Thus for two operands x and y and operator f, the syntax is fxy or $f(x, y)$. See also *functor, infix, postfix,* and *relation*.

prey-predator relation   Two biological species are in the following relation: the growth rate of the predator increases with increasing prey population, and the growth rate of the prey decreases with increasing predator population. The popular examples are fish and sharks, and deer and wolves.

primary sample   The collection of one of more increments or units initially taken from a population.

The potions may be either combined (composited or bulked sample) or kept separate (gross sample). If combined and mixed to homogeneity, it is a blended bulk sample. The term "bulk sample" is commonly used in the sampling literature as the sample formed by combining increments. The term is ambiguous since it could also mean a sample from a bulk lot, and it does not indicate whether the increments or units are kept separate or combined. Such use should be discouraged because less ambiguous alternative terms (composite sample, aggregate sample) are available. "Lot sample" and "batch sample" have also been used for this concept, but they are self-limiting terms. The use of "primary" in this sense is not meant to imply the necessity for multistage sampling.

primitive change   One of the conceptually simpler molecular changes into which an *elementary reaction* can be notionally dissected. Such changes include bond rupture, bond formation, internal rotation, change of bond length or bond angle, bond *migration*, redistribution of charge, etc.

The concept of *primitive change* is helpful in the detailed verbal description of elementary reactions, but a *primitive change* does not represent a process that is by itself necessarily observable as a component of an elementary reaction.

principal bundle   A bundle $(P, M, p; G)$ with a *Lie group G* as *standard fiber* and *transition functions* valued in the same G and acting on G by means of left translations $L_g : G \to G : h \mapsto g \cdot h$. On *principal bundles* a global and *canonical right action R_g* of G is defined by

$$R_g : P \to P : (x, h) \mapsto (x, h \cdot g).$$

This action is *free*, *transitive* on fibers and *vertical* (i.e., it is formed by vertical automorphisms), and it completely characterizes the *principal bundle*. An object is called *equivariant* if it preserves the canonical right action R_g.

On a *principal bundle* there is a one-to-one correspondence between local trivializations and local sections. Hence, a *principal bundle* is trivial if and only if it allows a global section.

principal connection   A connection over a *principal bundle* $(P, M, p; G)$ which preserves the canonical right action, i.e.,

$$T R_g(H_p) = H_{p \cdot g}.$$

In physics it is also called a *gauge field*.

product   In chemistry and biochemistry, the molecular species yielded by a reaction running in a particular direction.

Comment: The distinction is between the molecular result of a reaction and where a compound is represented in the arbitrarily written formal reaction equation. When a reaction occurs in a particular direction (either forward or backward), the *substrates* and *products* are known, and this identification of molecules with chemical roles is *independent* of where the molecules appear in the formal equation representing the

reversible reaction. See also *dextralateral, direction, dynamic equilibrium, formal reaction equation, microscopic reversibility, rate constant, reversibility, sinistralateral*, and *substrate*.

projectable vector field   A projectable vector field is a vector field on a *bundle* $(B, M, \pi; F)$ inducing by projection a vector field on the base *manifold M*. Locally, it is expressed as

$$\Xi = \xi^\mu(x)\partial_\mu + \xi^i(x, y)\partial_i$$

where $(x^\mu; y^i)$ are fibered coordinates. The flow of a projectable vector field is formed by *fibered automorphisms*. Vector fields that are not projectable do not preserve the bundle structure.

protein folding   The dynamical process of obtaining its structure (see protein structure), under appropriate conditions such as temperature, pH, etc., from an arbitrary initial structure. In test tubes, it is known experimentally that many proteins can reach their respective, almost unique three-dimensional structures. How this process works is still not completely known, even though it is generally accepted that the process is governed by the intramolecular interactions between atoms and interaction between the molecule and water in which the process occurs.

protein structure   A protein molecule has a well-defined three-dimensional structure in terms of the relative spatial coordinates of all its atoms. Most of our current knowledge about the protein structures are derived from laboratory determinations of the structures of many proteins using the methods of x-ray crystallography or nuclear magnetic resonance spectroscopy. The former requires a protein to form a crystal, while the latter can obtain the structure of a protein in aqueous solution. No reliable computational method exists yet for obtaining the spatial structure of a protein from its chemical structure (known as its amino acid sequence in biochemical literature).

pseudodifferential operator   A *pseudodifferential operator P* of order k on a compact *manifold M* is locally of the form: for any $u \in C_c^\infty(M)$

$$Pu(x) =$$
$$(2\pi)^{-n} \int \int e^{i(x-y)\cdot\xi} a(x, y, \xi) \times u(y) dy d\xi,$$

where $a(x, y, \xi)$ is a *symbol* of order k. This defines a bounded linear operator between the Sobolev spaces $P : H_c^s(M) \to H_c^{s-k}(M)$.

psychometry   The use of a wet-and-dry bulb thermometer for measurement of atmospheric humidity.

purine bases   Purine and its substitution derivatives, especially naturally occurring examples.

pyrimidine bases   Pyrimidine and its substitution derivatives, especially naturally occurring examples.

Q

QCD   See *chromodynamics, quantum.*

quantum yield, Φ   The number of defined events which occur per *photon* absorbed by the system. The integral quantum yield is:

Φ = (number of events)/(number of photons absorbed)

For a photochemical reaction:

Φ = (amount of reactant consumed or product formed)/(amount of photons absorbed)

$$\Phi = \frac{d[x]/dt}{n}$$

where $d[x]/dt$ is the rate of change of a measurable quantity, and n the amount of *photons* (mol or its equivalent einstein) absorbed per unit time. Φ can be used for photophysical processes or photochemical reactions.

quasi-optimal   Describing a V-conforming finite element scheme used to discretize a variational problem posed over the space V with the property that the discrete solutions u_h satisfy

$$\|u - u_h\|_V \le C \inf_{v_h \in V_h} \|u - v_h\|_V.$$

Here, $u \in V$ denotes the solution of the continuous variational problem. In a strict sense, the constant $C > 0$ must not depend on the choice of V_h. In a loose sense, it may depend on the family of finite elements, but not on the *triangulation*, on which V_h is built. In a very loose sense, if families of triangulations are involved, one demands that the constant C be independent of the *mesh width*, but it may depend on *shape regularity*.

quasi-uniform   A family $\{\Omega_h\}_{h \in \mathbb{H}}$, $\mathbb{H}$ an index set, of *triangulations* of a domain $\Omega \subset \mathbb{R}^n$ is called quasi-uniform, if there exists a $C > 0$ such that

$$\sup \left\{ \frac{\max\{\operatorname{diam}(K), K \in \Omega_h\}}{\min\{\operatorname{diam}(K), K \in \Omega_h\}}, h \in \mathbb{H} \right\} \le C.$$

Sloppily speaking, the cells of all meshes of a quasi-uniform family have about the same diameter. This permits us to provide asymptotic a priori estimates for interpolation and approximation errors of *finite element spaces* in terms of the *mesh widths*.

quotient space   See *equivalence relation.*

R

radiant (energy) flux, *P*, Φ   Although flux is generally used in the sense of the "rate of transfer of fluid, particles or energy across a given surface," the radiant energy flux has been adopted by IUPAC as equivalent to *radiant power, P*. ($P = \Phi = dQ/dt$, simplified expression: $P = \Phi = Q/t$ when the radiant energy, Q, is constant over the time considered). In photochemistry, Φ is reserved for *quantum yield*.

radiant exposure, *H*   The *irradiance, E* integrated over the time of irradiation ($\int E\,dt$, simplified expression $H = Et$ when the irradiance is constant over the time considered). For a parallel and perpendicularity, incident beam not scattered or reflected by the target or its surroundings, *fluence* (H_0) is an equivalent term.

radiant power, *P*, Φ   Same as *radiant (energy) flux*, Φ. Power emitted, transferred, or received as radiation.

radiation   A term embracing electromagnetic waves as well as fast moving particles. In *radioanalytical chemistry* the term usually refers to radiation emitted during a nuclear process (*radioactive decay*, nuclear reaction, *nuclear fission*, accelerators).

radical (free radical)   A *molecular entity* such as CH_3, SnH_3, $Cl^{\cdot}$ possessing an unpaired electron. (In these formulae the dot, symbolizing the unpaired electron, should be placed so as to indicate the atom of highest spin density, if this is possible.) Paramagnetic metal ions are not normally regarded as radicals. However, in the "isolobal analogy" the similarity between certain paramagnetic metal ions and radicals becomes apparent.

At least in the context of physical organic chemistry, it seems desirable to cease using the adjective "free" in the general name of this type of *chemical species* and molecular entity, so that the term "free radical" may in the future be restricted to those radicals which do not form parts of radical pairs.

Depending upon the core atom that possesses the unpaired electron, the radicals can be described as carbon-, oxygen-, nitrogen-, metal-centered radicals. If the unpaired electron occupies an orbital having considerable s or more or less pure p character, the respective radicals are termed σ- or π-radicals.

In the past, the term "radical" was used to designate a *substituent group* bound to a molecular entity, as opposed to "free radical," which nowadays is simply called radical. The bound entities may be called *groups* or substituents, but should no longer be called radicals.

radioactive   The property of a *nuclide* of undergoing spontaneous nuclear transformations with the emission of *radiation*.

radioactive decay   *Nuclear decay* in which particles or electromagnetic *radiation* are emitted or the nucleus undergoes spontaneous fission or electron capture.

radioactivity   The property of certain *nuclides* of showing *radioactive decay*.

radiochemistry   That part of chemistry which deals with *radioactive* materials. It includes the production of *radionuclides* and their compounds by processing irradiated materials or naturally occurring radioactive materials, the application of chemical techniques to nuclear studies, and the application of radioactivity to the investigation of chemical, biochemical, or biomedical problems.

radioluminescence   *Luminiscence* arising from excitation by high energy particles or radiation.

radionuclide   A *nuclide* that is *radioactive*.

radius of gyration, *s*   A parameter characterizing the size of a particle of any shape.

For a rigid particle consisting of mass elements of mass m_i, each located at a distance r_i from the center of mass, the radius of gyration s, is defined as the square root of the mass-average of the r_i^2 for all the mass elements, i.e.,

$$s = \left(\frac{\sum_i m_i r_i^2}{\sum_i m_i} \right)^{1/2}$$

For a nonrigid particle, an average overall conformation is considered, i.e.,

$$\langle s^2 \rangle^{1/2} = \frac{\langle \sum_i m_i r_i^2 \rangle^{1/2}}{\left(\sum_i m_i \right)^{1/2}}$$

The subscript zero is used to indicate unperturbed dimensions, as in $\langle s^2 \rangle_0^{1/2}$.

range   Let r be a relation with domain A and codomain B, and let $a \in A$, $r(a) \in B$. Then the *range* of r is the set of all images, $r(a_i)$, $\forall a_i \in A$. The range is denoted by either $r(A)$ or I.

Comment: See the comment on *relation* for more details. In particular, note that the range and *codomain* of a relation are not necessarily equivalent. See also *domain, image,* and *relation.*

rate coefficient   See *order of reaction* and *kinetic equivalence.*

rate constant   The parameter expressing the intrinsic rate at which a reaction can proceed in a particular direction; denoted k, usually with a subscript indicating which direction of the reaction is meant.

Comment: This parameter is also called the "intrinsic rate constant." For any reaction the quotient of the rate constants (forward over reverse) is the *equilibrium constant* for the reaction as written in a particular formal reaction equation. The actual rate achieved in a particular direction is the product of the rate constant and the concentrations of each of the obligatorily coreacting species for that direction. See also *dextralateral, direction, dynamic equilibrium, formal reaction equation, microscopic reversibility, product, reversibility, sinistralateral,* and *substrate.*

rate-controlling step   The rate-controlling (rate-determining or rate-limiting) step in a reaction occurring by a composite reaction sequence is an *elementary reaction*, the rate constant for

which exerts a strong effect, stronger than that of any other rate constant, on the overall rate. It is recommended that the expressions rate-controlling, rate-determining, and rate-limiting be regarded as synonymous, but some special meanings sometimes given to the last two expressions are considered under a separate heading.

A rate-controlling step can be formally defined on the basis of a control function (or control factor) CF, identified for an elementary reaction having a rate constant k_i by

$$CF = (\partial \ln v \partial \ln k_i) K_j, k_j$$

where v is the overall rate of reaction. In performing the partial differentiation all equilibrium constants K_j and all rate constants except k_i are held constant. The elementary reaction having the largest control factor exerts the strong influence on the rate v and a step having a CF much larger than any other step may be said to be rate-controlling.

A rate-controlling step defined in the way recommended here has the advantage that it is directly related to the interpretion of *kinetic isotope effects.*

As formulated, this implies that all rate constants are of the same dimensionality. Consider, however, the reaction of A and B to give an intermediate C, which then reacts further with D to give products

$$A + B \underset{k_{-1}}{\overset{k_1}{\rightleftarrows}} C \qquad (1)$$

$$C + D \xrightarrow{k_2} \text{Products.} \qquad (2)$$

Assuming that C reaches a *steady state*, then the observed rate is given by

$$v = \frac{k_1 k_2 [A][B][D]}{k_{-1} + k_2[D]}.$$

Considering $k_2[D]$ a pseudo-first-order rate constant, then $k_2[D] >> k_{-1}$, and the observed rate $v = k_1[A][B]$ and $k_{obs} = k_1$. Step (1) is said to be the rate-controlling step.

If $k_2[D] << k_{-1}$, then the observed rate

$$v = \frac{k_1 k_2}{k_{-1}} [A][B][D].$$

rate law (empirical differential rate equation)
An expression for the *rate of reaction* of a particular reaction in terms of concentrations of *chemical species* and constant parameters (normally *rate coefficients* and partial *orders of reaction*) only. For examples of rate laws see equations (1)-(3) under *kinetic equivalence*.

rate of change (of a function $f(t)$, with respect to t) (**1**) *Average rate of change* over $[t_1, t_2]$ is

$$\frac{f(t_2) - f(t_1)}{t_2 - t_1}.$$

(**2**) *Instantaneous rate of change* at $t = t_1$ is

$$\frac{df}{dt}(t_1).$$

reactant A substance that is consumed in the course of a chemical reaction. It is sometimes known, especially in the older literature, as a reagent, but this term is better used in a more specialized sense as a test substance that is added to a system in order to bring about a reaction or to see whether a reaction occurs (e.g., an analytical reagent).

reaction Any chemical or biochemical transformation of matter and energy.
Comment: This definition includes both spontaneous and catalyzed reactions. The word "transformation" is used in the general English sense, not the mathematical or linguistic sense.

reaction coordinate A geometric parameter that changes during the conversion of one (or more) reactant *molecular entities* into one (or more) product *molecular entities* and whose value can be taken for a measure of the progress of an *elementary reaction* (for example, a bond length or bond angle or a combination of bond lengths and/or bond angles; it is sometimes approximated by a nongeometric parameter, such as the *bond order* of some specified bond). In the formalism of "*transition-state theory*," the reaction coordinate is that coordinate in a set of curvilinear coordinates obtained from the conventional ones for the reactants which, for each reaction step, leads smoothly from the configuration of the reactants through that of the transition state to the configuration of the products. The reaction coordinate is typically chosen to follow the path along the gradient (path of shallowest ascent/deepest descent) of potential energy from reactants to products.

The term has also been used interchangeably with the term transition coordinate, applicable to the coordinate in the immediate vicinity of the potential energy maximum. Being more specific, the name transition coordinate is to be preferred in that context.

reaction-diffusion equation Let $U \subset \mathbb{R}^n$ be open and $u : U \times \mathbb{R} \to \mathbb{R}^n$. The *reaction-diffusion equation* for u is

$$u_t - \Delta u = f(u).$$

reaction path (**1**) A synonym for *mechanism*.
(**2**) A trajectory on the *potential-energy surface*.
(**3**) A sequence of synthetic steps.

reaction's compounds The set of molecular species, including catalysts, which participate in that reaction. See *compound's reactions*.

reactive (reactivity) As applied to a *chemical species*, the term expresses a kinetic property. A species is said to be more reactive or to have a higher reactivity in some given context than some other (reference) species if it has a larger rate constant for a specified *elementary reaction*. The term has meaning only by reference to some explicitly stated or implicitly assumed set of conditions. It is not to be used for reactions or reaction patterns of compounds in general.

The term is also more loosely used as a phenomenological description not restricted to elementary reactions. When applied in this sense the property under consideration may reflect not only rate, but also equilibrium, constants.

reduced viscosity (of a polymer) The ratio of the *relative viscosity increment* to the mass concentration of the *polymer*, c, i.e., η_i/c, where η_i is the relative viscosity increment.
Notes: (**1**) The unit must be specified; cm^3g^{-1} is recommended.
(**2**) This quantity is neither a viscosity nor a pure number. The term is to be looked on as

a traditional name. Any replacement by consistent terminology would produce unnecessary confusion in the polymer literature. Synonymous with viscosity number.

reference material   A substance or mixture of substances, the composition of which is known within specified limits, and one or more of the properties of which is sufficiently well established to be used for the calibration of an apparatus, the assessment of a measuring method or for assigning values to materials. Reference materials are available from national laboratories in many countries [e.g., National Institute for Standards and Technology (NIST), U.S., Community Bureau of Reference, U.K.].

regulatory motif   A conserved *topological motif* which reduces or increases the activity of a gene, enzyme, or *macromolecular* complex.

Comment: See also *biochemical, chemical, dynamical, functional, kinetic, mechanistic, phylogenetic, thermodynamic,* and *topological motives.*

relation (between two sets)   A relation between A and B is any subset $R \subset A \times B$ of the Cartesian product of the two sets. If $(a, b) \in R$ we say that *a is related to b* (in that order), and we write $a\ R\ b$. If R can be produced by some operation or procedure r, then we often call r the relation.

For example, consider $A = \{a, b, c\}$ and $B = \{1, 2, 3\}$, and let $r(a_i, b_i), a_i \in A, b_i \in B, 1 \leq i \leq 3$. Then $R = \{(a, 1), (b, 2), (c, 3)\}$.

Consider a relation r between two sets A and B. A is called the *domain* of r and B its *codomain*. The elements of B related, via a relation r to an element $a \in A$ is called the *image* of a under r, and denoted by $r(a)$. The union $r(A)$ of all $r(a), a \in A$, is called the *range* of r, and can also be denoted I. Thus, $r(A)$ may be a proper subset of B. So, the range is not necessarily the same as the codomain.

See also *domain, image,* and *range,* and *relation.*

A relation f is called a function if for each $a \in A$ there exists a unique $b \in B$ such that $a\ f\ b$. If f is a function and $a\ f\ b$ we write $f(a) = b$ and $f : A \to B$, with $f : a \mapsto b$. Notice that according to this definition, which is standard in algebra, the map $f : x \mapsto \frac{1}{x}$ is not a function $f : \mathbb{R} \to \mathbb{R}$ since it is not defined in $0 \in \mathbb{R}$. Of course, it is a function on the smaller space $\mathbb{R} - \{0\}$.

relative molar mass   Molar mass divided by 1 g mol^{-1} (the latter is sometimes called the standard molar mass).

relative molecular mass, M_r   Ratio of the mass of a molecule to the *unified atomic mass unit.* Sometimes called the molecular weight or *relative molar mass.*

relative responsivity   See *responsivity.*

relative viscosity   The ratio of the viscosity η of the solution to the viscosity η_s of the solvent, i.e., $\eta_r = \eta/\eta_s$. Synonymous with viscosity ratio.

relative viscosity increment   The ratio of the difference between the viscosities of solution and solvent to the viscosity of the solvent, i.e., $\eta_i = (\eta - \eta_s)/\eta_s$, where η is the viscosity of the solution and η_s is the viscosity of the solvent.

The use of the term "specific viscosity" for this quantity is discouraged, since the relative viscosity increment does not have the attributes of a specific quantity.

relaxation oscillator   For some oscillatory dynamics with a limit cycle, parts of the cycle are traversed quickly in comparison with other parts. This is often referred to as a *relaxation oscillator.* This phenomenon suggests that the underlying ordinary differential equations for such a system have a small parameter which is present in a crucial place to cause this rapid variation in the solution. A widely used class of models for such phenomena is a system $d\mathbf{u}/dt = \mathbf{f}(\mathbf{u}, \alpha)$, $d\alpha/dt = \epsilon g(\mathbf{u}, \alpha)$ which can be analyzed by singular perturbation for small ϵ.

rendering   A physical model or drawing of an object intended for direct perception by humans.

Comment: The key notion is that the model is directly perceived by humans, usually visually. For example, renderings can be images drawn on a raster screen, a physical molecular model, or a projection in a virtual reality environment. A rendering is distinct from the information it contains or its abstract specification, and is strongly determined by the device used for its achievement and the intended method of perception. Thus renderings of the same landscape in oils and water colors can have distinctly different characters, even if the landscapes are completely *isomorphic.* Synonymous with *drawing.*

representation   (**1**) A *representation* of an object $a \in \mathcal{A}$ is an image $\psi_q(a) \in \Psi$ which preserves a property q such that

(i.) *uniqueness:* the mapping, ψ_q, between $\mathcal{A}$ and Ψ for q is one-to-one and onto (notated $\psi_q : a \longmapsto \psi_q(a), a \in \mathcal{A}, \psi_q(a) \in \Psi$, and

(ii.) *equivalent transformations:* for some transformation ζ over the set of objects $\mathcal{A}$,

$$\zeta : a \longmapsto a',$$

$a, a' \in \mathcal{A}$, there exists a transformation ζ' over the set of representations,

$$\zeta' : \psi_q(a) \longmapsto \psi_q(a'),$$

$\psi_q(a), \psi_q(a') \in \Psi$, such that the result of ψ_q applied to $\zeta(a)$ is identical to the result of ζ' applied to $\psi_q(a)$, or

$$\psi_q \circ \zeta(a) = \zeta' \circ \psi_q(a) = \psi_q(a')$$

where $\circ$ is the composition operator. $\psi_q(\zeta) = \zeta'$, so we say that ζ and ζ' are *equivalent* transformations.

Equally, a set of such representations Ψ.

Comment: The intent here is to capture two of the most salient features of a representation as the term is commonly used in artificial intelligence and databases. The first is that the representation is simply an image of something in the "real world" which preserves some property important to the user, geometric isomerism, hair color, or reaction rate. In general expressive representations will mirror the object closely and in ways that reflect how humans conventionally represent that object in their heads, often *via* language. However, it must be remembered that strictly speaking, there is no requirement for any relationship between a term and a representation for an object, apart from the properties of the transformations between them. Note that a database can contain multiple representations which upon (internal) transformation are informationally equivalent, provided that each preserves a different property. Thus a molecule can be represented by a configuration rule, a key-pair list, or the terminal form. Each preserves a different property of the molecule (configuration of substituent groupings, edges in the structural graph, atoms, and bonds); all are interconvertible by the grammar; and each is optimized for a specific class of computation.

The second is that if the representation is to be useful for computations which do something more sophisticated than simply looking up data, it must permit one to define a computational transformation which mimics a "real-world operation" sufficiently so that for the preserved property of interest, the result of the computational transformation is the image of the result of the real-world operation. The word is used as both a singular and a collective noun. See also *semantics* and *semiote.*

(**2**) (of a group G) A linear *left action* of a group on a *vector space V.* To any group element $g \in G$ an *automorphism* of V is associated. Example, the group of *rotations* SO(3) is represented on $\mathbb{R}^3$ by matrix multiplication.

representative sample   A sample resulting from a sampling plan that can be expected to reflect adequately the properties of interest of the parent population.

A representative sample may be a random sample or, for example, a stratified sample, depending upon the objective of sampling and the characteristics of the population. The degree of representativeness of the sample may be limited by cost or convenience.

111

resolvent   See *spectrum*.

resonance effect   See *mesomeric effect*.

responsivity (in detection of radiation), **R**
Detector input can be, e.g., radiant power, irradiation, or radiant energy. It produces a measurable detector output which may be, e.g., an electrical charge, an electrical current or potential or a change in pressure. The ratio of the detector output to the detector input is defined as the responsivity. It is given in, e.g., ampere/watt, volt/watt. The *responsivity* is a special case of the general term *sensitivity*.

Dark current is the term for the electrical output of a detector in the absence of input. This is a special case of the general term *dark output*. For photoconductive detectors the term *dark resistance* is used.

If the responsivity is normalized with regard to that obtained from a reference radiation, the resulting ratio is called *relative responsivity*. For measurements with monochromatic radiation at a given wavelength λ the term spectral responsivity $R(\lambda)$ is used. In some cases the relative spectral responsivity, where the spectral responsivity is normalized with respect to the *responsivity* at some given wavelength, is used. The dependence of the spectral responsivity on the wavelength is described by the *spectral responsivity function*. The useful spectral range of the detector should be given as the wavelength range where the relative responsivity does not fall below a specified value.

rest point (of a balance)   The position of the pointer with respect to the pointer scale when the motion of the beam has ceased.

retinoids   Oxygenated derivatives of 3,7-dimethyl-1-(2,6,6-trimethylcyclohex-1-enyl) nona-1,3,5,7-tetraene and derivatives thereof.

reversibility   In chemistry and biochemistry, the notion that any reaction can proceed in both directions, at least to some extent. See also *dextralateral, direction, dynamic equilibrium, formal reaction equation, microscopic reversibility, product, rate constant, sinistralateral*, and *substrate*.

Ricci scalar   On a *manifold M* with a connection Γ and a metric g the contraction of the *Ricci tensor* of Γ with the inverse metric

$$R = g^{\mu\nu} R_{\mu\nu}.$$

Ricci tensor   The contraction of the *Riemann tensor* defined by

$$R_{\beta\nu} = R^{\alpha}_{\cdot\,\beta\alpha\nu} = -R^{\alpha}_{\cdot\,\beta\nu\alpha}.$$

It is symmetric in the indices (β, ν).

Riemannian manifold   A pair (M, g) formed by a *manifold M* of dimension m and a Riemannian metric g on it. If the metric is positive definite (i.e., of signature $(m, 0)$) then it is called *strictly Riemanniann*; if the metric is indefinite then it is called *pseudo-Riemannian*.

Riemannian metric   A positive definite inner product on the tangent space to a manifold at x, for each point x of the manifold, varying continuously with x.

right action (of a group on a space X)   A map $\rho : X \times G \to X$ such that:

   (i.) $\rho(x, e) = x$;
   (ii.) $\rho(x, g_1 \cdot g_2) = \rho(\rho(x, g_1), g_2)$;

where G is a group, e its neutral element, · the product operation in G, and X a topological space. The maps $\rho_g : X \to X$ defined by $\rho_g(x) = \rho(x, g)$ are required to be *homeomorphisms*. If X has a further structure one usually requires ρ to preserve the structure. If ρ is a right action, then one can define a *left action* by setting $\lambda(g, x) = \rho(x, g^{-1})$. See also *left action*.

right invariance   The property that an object on a *manifold M* is invariant with respect to a *right action* of a group on M. For example, a vector field $X \in \mathfrak{X}(M)$ is right invariant with respect to the right action $\rho_g : M \to M$ if and only if:

$$T_x\rho_g\,X(x) = X(x \cdot g).$$

right translations (on a group G)   The *right action* of G onto itself defined by $\rho(h, g) =$

$\rho_g(h) = h \cdot g$. Notice that, if G is a *Lie group*, $\rho_g \in \text{Diff}(G)$ is a *diffeomorphism* but not a *homomorphism* of the group structure. See also *left translations* and *adjoint representations*.

ring   A nonempty set R, with two binary operations $+$ and $\cdot$, which satisfy the following axioms:

    (i.)   with respect to $+$, R is an *Abelian group*;

    (ii.) $\cdot$ is associative: $a \cdot (b \cdot c) = (a \cdot b) \cdot c$ for all $a, b, c \in R$;

    (iii.) $+$ and $\cdot$ satisfy the distributive laws: $a \cdot (b + c) = a \cdot b + a \cdot c$ and $(b + c) \cdot a = b \cdot a + c \cdot a$, for $a, b, c \in R$. R is called a *commutative ring* if it is commutative with respect to $\cdot$.

rotation   An *orthogonal tranformation* of a Euclidean space (V, g). See *orthogonal group*.

S

saddle point   Let $f : X \times Y \longrightarrow R$. Then, (x^*, y^*) is a *saddle point* of f if x^* minimizes $f(x, y^*)$ on X, and y^* maximizes $f(x^*, y)$ on Y. Equivalently,

$$f(x^*, y) \leq f(x^*, y^*) \leq f(x, y^*)$$

for all x in X, y in Y.

von Neumann (1928) proved this equivalent to:

$$\text{Inf } [\text{Sup}\{f(x, y) : y \in Y\} : x \in X]$$

$$= \text{Sup}[\text{Inf}\{f(x, y) : x \text{ in } X\} : y \in Y]$$

$$= f(x^*, y^*).$$

saddle point problem   On Banach spaces V, W consider the functional $J : V \times W \to \mathbb{R}$. A *saddle point problem* seeks a pair $(u, p) \in V \times W$ such that

$$J(u, p) = \inf_{v \in V} \sup_{q \in W} J(v, q).$$

Necessary conditions for this kind of stationary point often lead to symmetric linear variational problems with saddle point structure: seek $u \in V$, $p \in W$ such that

$$a(u, v) + b(v, p) = f(v) \ \forall v \in V,$$

$$b(u, q) = g(q) \ \forall q \in W,$$

where $a : V \times V \to \mathbb{C}, b : V \times W \to \mathbb{C}$ are *sesqui-linear* forms, and f, g stand for linear forms on V and W, respectively. Specimens of variational *saddle point problems* are encountered in the case of *mixed variational formulations*, the Stokes problem of fluid mechanics, and whenever a linear constraint is taken into account by a Lagrangian multiplier.

sample (in analytical chemistry)   A portion of material selected from a larger quantity of material. The term needs to be qualified, e.g., *bulk sample, representative sample, primary sample*, bulked sample, or *test sample*.

The term "sample" implies the existence of a sampling error, i.e., the results obtained on the portions taken are only estimates of the concentration of a constituent or the quantity of a property present in the parent material. If there is no or negligible sampling error, the portion removed is a test portion, aliquot, or specimen. The term "specimen" is used to denote a portion taken under conditions such that the sampling variability cannot be assessed (usually because the population is changing), and is assumed for convenience, to be zero. The manner of selection of the sample should be prescribed in a sampling plan.

sample unit   The discrete identifiable portion suitable for taking as a sample or as a portion of a sample. These units may be different at different stages of sampling.

satisfiability problem   Find a truth assignment to logical propositions such that a (given) collection of clauses is true (or ascertain that at least one clause must be false in every truth assignment). This fundamental problem in computational logic forms the foundation for NP-completeness

scalar   Given a vector space V, a member of the field from which scalar multiplication of vectors in V is defined.

scalar product   See *inner product*.

scaling   Changing the units of measurement, usually for the numerical stability of an algorithm. The variables are transformed as $x' = Sx$, where $S = \text{diag}(s_j)$. The diagonal elements are the *scale values*, which are positive: $s_1, \ldots, s_n > 0$. Constraint function values can also be scaled. For example, in an LP, the constraints $Ax = b$, can be scaled by $RAx = Rb$, where $R = \text{diag}(r_i)$ such that $r > 0$. (This affects the dual values.) Some LP scaling methods simply scale each column of A by dividing by its greatest magnitude (null columns are identified and removed).

Example	A column scaling	A row scaling
10x+100y=500	.333x+y=500	x+10y =50
30x +.3y =.2	x+.003y =.2	300x +3y =2

Another method is *logarithmic scaling*, which scales by the logarithm of the greatest magnitude. More sophisticated methods are algorithmic, taking both row and column extremes into account.

scaling argument (homogeneity argument) The idea is to exploit the behavior of norms of functions or vector fields under pull-backs related to scaling transformations $\mathbf{x} \mapsto \delta\mathbf{x}$, $\delta > 0$. Very closely related to techniques relying on *parametric equivalence*.

scatter search A population-based meta-heuristic, starting with a collection of reference points, usually obtained by the application of some heuristic. A new point is created by taking combinations of points in the population, and rounding elements that must be integer valued. It bears some relation to a genetic algorithm, except that scatter search uses linear combinations of the population, while the GA crossover operation can be nonlinear.

scheduling (e.g., jobs) A schedule for a sequence of jobs, say $j_1, \ldots, j_n$, is a specification of start times, say $t_1, \ldots, t_n$, such that certain constraints are met. A schedule is sought that minimizes cost and/or some measure of time, like the overall project *completion time* (when the last job is finished) or the *tardy time* (amount by which the completion time exceeds a given deadline). There are precedence constraints, such as in the construction industry, where a wall cannot be erected until the foundation is laid.

There is a variety of scheduling heuristics. Two of these for scheduling jobs on machines are list heuristics: the *Shortest Processing Time* (SPT) and the *Longest Processing Time* (LPT). These rules put jobs on the list in non-decreasing and non-increasing order of processing time, respectively.

Other scheduling problems, which might not involve sequencing jobs, arise in production planning.

Schrödinger equation Let $U \subset \mathbb{R}^n$ be open and $u : U \times \mathbb{R} \to \mathbb{R}$. The *Schrödinger equation* for u is

$$iu_t + \Delta u = 0.$$

scintillators Materials used for the measurement of *radioactivity*, by recording the *radioluminescence*. They contain compounds (*chromophores*) which combine a high fluorescence quantum efficiency, a short fluorescence lifetime, and a high solubility. These compounds are employed as solutes in aromatic liquids and *polymers* to form organic liquid and plastic scintillators, respectively.

search tree The tree formed by a branch and bound *algorithm* strategy. It is a tree because at each (forward) branching step the problem is partitioned into a disjunction. A common one is to dichotomize the value of some variable, $x \leq v$ or $x \geq v + 1$. This creates two nodes from the parent:

[parent node]

$x \leq v$ $x \geq v + 1$

[left child]   [right child]

secant The function

$$\sec(x) = \frac{1}{\cos x}$$

See *cosine*.

secant method A method to find a root of a univariate function, say F. The iterate is

$$x^{(k+1)} = x^k - \frac{F(x^k)[x^k - x^{(k-1)}]}{F(x^k) - F(x^{(k-1)})}.$$

If F is in C^2 and $F''(x) \neq 0$, the order of convergence is the golden mean, say, g (approx. $=$ 1.618), and the limiting ratio is:

$$\left| \frac{2F'(x)}{F''(x)} \right|^{(g-1)}.$$

second-order conditions A descendant from classical optimization, using the second-order term in Taylor's expansion. For unconstrained optimization, the second-order necessary condition (for f in C^2) is that the Hessian is negative semidefinite (for a max). Second-order sufficient conditions are the first-order conditions plus the requirement that the Hessian be negative definite.

For constrained optimization, the second-order conditions are similar, using projection for a regular mathematical program and the Lagrange multiplier rule. They are as follows (all functions are in C^2, and the mathematical program is in standard form, for x^* a local maximum):

(i.) *Second-order necessary conditions.* There exist Lagrange multipliers, (u, v), such that $u \geq 0$ and $u g(x^*) = 0$ for which: (1) $\text{grad}_x[L(x^*, u, v)] = 0$, and (2) $H_x[L(x^*, u, v)]$ is negative semi-definite on the tangent plane.

(ii.) *Second-order sufficient conditions.* The above necessary conditions hold but with (2) replaced by ($2'$) $H_x[L(x^*, u, v)]$ is negative definite on the tangent plane.

selectively labeled An *isotopically labeled* compound is designated as selectively labeled when a mixture of *isotopically substituted* compounds is formally added to the analogous *isotopically unmodified* compound in such a way that the position(s) but not necessarily the number of each labeling *nuclide* is defined. A selectively labeled compound may be considered as a mixture of *specifically labeled* compounds.

self-adjoint operator A *linear* operator T on a *Hilbert space* $(H, <, >)$ such that $T^* = T$, where the (Hilbert)-*adjoint* T^* is defined by $< x, Ty >=< T^*x, y >$, $x, y \in H$. See also *symmetric operator*. A symmetric operator T is called *essentially self-adjoint* if its closure $\bar{T}$ is self-adjoint.

self-avoiding random walk A random walk which does not pass any space point twice. In three dimensions, this is a more realistic model for *polymer* chains. See *Gaussian chain* and *excluded volume*.

self-concordance Properties of a function that yield nice performance of Newton's method used for line search when optimizing a barrier function. Specifically, let B be a barrier function for $S = \{x \in X : g(x) \leq 0\}$ with strict interior S^0. Let x be in S and let d be a direction vector in $\mathbb{R}^n$ such that the line segment $[x - td, x + td]$ is in S for t in $[0, t^*]$, where $t^* > 0$. Then, define $F : [0, t^*] \longrightarrow \mathbb{R}$ by:

$$F(t) = B(x + td)$$

(while noting that F depends on x and d). The function F is *self-concordant* if it is convex in C^3 and satisfies the following for all x and d:

$$|F'''(0)| \leq 2F''(0)^{(3/2)}.$$

One calls F *k-self-concordant* in an open convex domain if

$$|F'''(0)| \leq 2k F''(0)^{(3/2)}.$$

The logarithmic barrier function, associated with linear programming, is self-concordant with $k = 1$. This further extends naturally to functions in $\mathbb{R}^n$.

semantic mapping A *bijective*, partial function between each member of the set of symbols, $\sigma_j \in \Sigma$, and its semantics: $\omega \colon \sigma_j \longmapsto \omega(\sigma_j)$, ω the mapping operator. The set of *semantic mappings*, Ω, is defined for each element of the domain and codomain to which they apply.

Comment: Notice that ω is a partial function. There will be elements of the domain (the symbol set of the language) for which a given mapping will not be defined. See also *semantics* and *semiote*.

semantics For each symbol σ_j in the alphabet Σ, j a positive integer index, its *semantics*, $\omega(\sigma_j)$, is a computationally executable definition of the meaning of σ_j. It is found or produced by applying a *semantic mapping* ω to σ_j, denoted $\omega \colon \sigma_j \longmapsto \omega(\sigma_j)$, such that ω is one-to-one, onto, and defined for that σ_j. Under these conditions, we call both symbol and mapping *semantically well formed*.

Comment: This is equivalent to saying that every term in a language, $\mathcal{L}$, which describes a *universe of discourse*, has a unique and precise meaning. This definition, which is appropriate for computation but not for linguistics, eliminates multiple connotations for a single term except as it can be recursively fragmented to other terms having unique semantics. In that case, however, the ultimate terms would necessarily be used in preference to the more connotation-rich term. The language can be formal or not, though it is more likely that any arbitrary mapping will be unique if it is derived from a formal language. Of course, natural language is used to describe the phenomenal universe and the objects within it (see *representation*); the meanings of terms in the natural language are mental constructs. Thus in defining the natural language term "reaction" corresponding to a biochemical transformation (the object), one specifies the type of information understood — chemical mechanism, kinetic regime, formal equation, etc., fragmenting the term into a set of terms, then mapping each term to a meaning. The representation of the object in the database is arbitrary and independent of the term referencing that object. Semantics are distinct from database schemata, which describe the relationships among objects internal to the database but depend on the observer to recognize the mappings. This notion of semantics echoes that of conceptual graphs, but without depending upon its graphical apparatus. See also *representation, semiote*, and *term semantics*.

semidefinite program   Min $\{cx : S(x) \in P\}$, where P is the class of positive semidefinite matrices, and $S(x) = S_0 + \text{Sum}_j\{x(j)S_j\}$, where each S_j, for $j = 0, \ldots, n$ is a (given) symmetric matrix. This includes the linear program as a special case.

semi-infinite program   A mathematical program with a finite number of variables or constraints, but an infinite number of constraints or variables, respectively. The randomized program is a semi-infinite program because it has an infinite number of variables when X is not finite.

semiote   A *semiote* is a symbol denoting the semantics of a useful elementary part of an idea, datum, or computation.

More formally, let a symbol be denoted σ_0 and the set of all symbols other than σ_0 be denoted

$\Sigma' = \Sigma - \{\sigma_0\}$. A member of the set Σ' is denoted σ_j, j a positive integer indexing Σ'. Then σ_0 is considered to be an *elementary symbol*, or *semiote*, if three conditions are fulfilled.

(i.) *well-formed*. There is one and only one well-formed mapping ω such that $\omega: \sigma_0 \longmapsto \omega(\sigma_0)$.

(ii.) *unique*. The symbol σ_0 and its semantics, $\omega(\sigma_0)$, are unique, or

$$\sigma_0 \neq \sigma_j, \ \omega(\sigma_0) \neq \omega(\sigma_j),$$

$\forall \, \sigma_j \in \Sigma', \omega(\sigma_j) \in \Omega(\Sigma')$.

(iii.) *elementary*. Denote the set of all possible constructs of symbols in Σ' by $\mathcal{C}'$, and a particular construct, c_k, k a positive integer index. Then σ_0 is elementary if, for a well-formed mapping ω, the semantics of every construct, $\omega(c_k)$, is not equal to the semantics of σ_0, $\omega(\sigma_0)$, or

$$\omega(\sigma_0) \neq \omega(c_k), \ \forall \, c_k \in \mathcal{C}'.$$

The semiote σ_0 is denoted ς. Every semiote has four properties:

(i.) its formally defined, computable semantics;

(ii.) its formally defined, computable syntax;

(iii.) its informally defined, natural language semantics; and

(iv.) its informally defined, natural language syntax.

The set of all *semiotes* is denoted Σ_ς, and is also called the *semantic basis set*. A construct of *semiotes* is called a *bundle of semiotes* or a *semiotic bundle*.

Comment: For the natural world there are as many mental models and domain models as there are scientists and databases. Some of their terms' semantics and representations will be isomorphic among people and databases (and between people and databases), but many will not. *Semiotes*, singly or more usually combined, are intended to map between the multiple meanings humans assign to terms of a language describing the phenomenal world and the multiple ways in which objects and operations on them from that world can be represented in databases. The proposition is that an abstract layer of *semiotes*, incrementally and distributively formulated and maintained, clearly specifies the semantics of these models for the purposes of exchanging computations. Each site offering data or computations to the network would locally maintain a publicly readable list of mappings between its internal representation and the semiotes and a parser implementing those mappings as needed by the local engine. Similarly, sites posing computations would maintain mappings and parser between their internal representation and the semiotes. All sites are free to change their schema, engines, and proffered services at any time in any manner. Flexibility of use and ease of definition is promoted by making semiotes representing the smallest useful part, and combining these together. See also *representation* and *semantics*.

sensitivity (**1**) (in mass spectrometry) Two different measures of sensitivity are recommended. The first, which is suitable for relatively involatile materials as well as gases, depends upon the observed change in ion current for a particular amount or change of flow rate of sample through the ion source. A second method of stating sensitivity, that is most suitable for gases, depends upon the change of ion current related to the change of partial pressure of the sample in the ion source.

It is important that the relevant experimental conditions corresponding to sensitivity measurement should always be stated. These include in a typical case details of the instrument type, bombarding electron current, slit dimensions, angular collimation, gain of the multiplier detector, scan speed, and whether the measured signal corresponds to a single mass peak or to the ion beam integrated over all masses. Some indication of the time involved in the determination should be given, e.g., counting time or bandwidth. The sensitivity should be differentiated from the detection limit.

(**2**) (in metrology and analytical chemistry) The slope of the *calibration curve*. If the *curve* is in fact a *"curve,"* rather than a straight line, then of course sensitivity will be a function of analytic concentration or amount. If *sensitivity* is to be a unique performance characteristic, it must depend only on the *chemical measurement process*, not upon scale factors.

sensitivity analysis The concern with how the solution changes if some changes are made in either the data or in some of the solution values (by fixing their value). *Marginal analysis*, which is concerned with the effects of small perturbations, may be measurable by derivatives. Parametric analysis is concerned with larger changes in parameter values that affect the data in the mathematical program, such as a cost coefficiency or resource limit.

Under suitable assumptions, the multipliers in the Lagrange multiplier rule provide derivatives of the optimal response function, i.e., under certain conditions, $(u, v) = \text{grad}^* f(b, c)$.

separable program The functions are separable: $f(x) = \sum_j f_j(x_j)$, $g(x) = \sum_j g_j(x_j)$, and $h(x) = \sum_j h_j(x_j)$. The classical (LP) approaches to separable programming are called *lambda-form* and *delta-form*, both using piecewise linear approximations.

Let $\{x^k\}$ be a specified set of points, where $x^k = [x(k, 1), x(k, 2), \ldots, x(k, n)]$, and let $y = \{y(k, j)\}$ be decision variables that are the coefficients of convex combinations, giving the following linear program:

$$\text{Max Sum}_k j\{y(k, j) f_j(x(k, j))\} : y \geq 0,$$

$$\sum_k y(k, j) = 1 \text{ for each } j,$$

$$\sum_{kj} y(k, j) g_j(x(k, j)) \leq 0,$$

$$\sum_{kj} y(k, j) h_j(x(k, j)) = 0.$$

A restricted basis entry rule is invoked during the simplex method to yield an approximate solution. (However, this is dominated by the generalized Lagrange multiplier method, which can be viewed as generating the approximating breakpoints *a posteriori*, getting successively finer near the solution.)

The delta form uses the differences: $u(k, j) = x(k, j) - x(k - 1, j)$. The associated functional differences are:

$$Df(k, j) = f_j(x(k, j)) - f_j(x(k - 1, j)),$$
$$Dg(k, j) = g_j(x(k, j)) - g_j(x(k - 1, j)),$$
$$Dh(k, j) = h_j(x(k, j)) - h_j(x(k - 1, j)).$$

Then, the approximating LP is:

$$\text{Max} \sum_{kj} Df(k, j)u(k, j) : 0 \leq u \leq 1,$$

$$\sum_{kj} Dg(k, j)u(k, j) \leq b,$$

$$\sum_{kj} Dh(k, j)u(k, j) = c,$$

where $b = -\sum_j g_j(x(0, j))$ and $c = -\sum_j h_j$ $(x(0, j))$ (a similar constant was dropped from the objective). Another restricted basis rule is invoked: $u(k, j) > 0$ implies $u(k, q) = 1$ for all $q < j$ and all k.

separating hyperplane   A hyperplane for which two (given) sets lie in opposite half spaces. The separation is *strict* if the two sets are contained in their respective open half space.

sequence   An ordered countable collection of elements which can include duplicates. The ordering need not reflect a mathematical relation. An enumerated sequence is delimited by angle brackets ($\langle x \rangle$).

Comment: The key notion of a sequence is that of succession: that for any two elements of a sequence, a and b, $a \prec b$ iff a occurs to the left of b in the sequence (we adopt the convention that we read the sequence from left to right). Be careful not to confuse $\prec$ with $<$; a could be 12 and b could be 5, yet still $a \prec b$ (read "a precedes b"). Classic examples of sequences are paths through graphs and strings, such as DNA sequences.

sequencing problems   Finding an ordering, or permutation, of a finite collection of objects, like jobs, that satisfies certain conditions, such as precedence constraints.

sequential decision process   See *time-staged*.

sequential linear programming (SLP)   Solving a nonlinear program by a sequence of linear approximations and using linear programming to solve each one. The linear approximations are usually done by using the first-order Taylor expansion.

sequential quadratic programming (SQP)   Solving a nonlinear program by a sequence of quadratic approximations and using quadratic programming to solve each one. The approximations are usually done by using the second-order Taylor expansion.

sequential unconstrained minimization technique (SUMT)   This is the penalty function approach.

sesqui-linear   Describing a complex-valued function of two variables which is linear in the first variable and conjugate-linear in the second.

series   A formal sum $\sum_j a_j$, where $< a_j >$ is a *sequence*.

set   An unordered collection of elements, without duplicates, each of which elements satisfies some property. An enumerated set is delimited by braces ($\{x\}$).

Comment: Some authors permit sets to include duplicates. See also *bag, list, sequence,* and *tuple*.

set difference   Given two sets, A and B, their *difference, $D = A - B$*, is the set of all elements of A not found in B: $\forall d \in D, d \in A$ and $d \notin B$.

Comment: Synonymous with *set subtraction*. Similar operations can be defined for bags, lists, and sequences. See also *symmetric difference*.

set of reactions   A collection of reactions.

set subtraction   See *set difference*.

shadow price   An economic term to denote the rate at which the optimal value changes with respect to a change in some right-hand side that represents a resource supply or demand requirement. This is sometimes taken as synonymous with the dual price, but this can be erroneous, as in the presence of degeneracy.

shape function   Given a *finite element* (K, V_K, Π_K) the set of local *shape functions* is the basis of V_K that is dual to Π_K.

shape regular   Let $\Omega_h, h \in \mathbb{H}$, $\mathbb{H}$ an index set, be a family of *meshes* of a domain $\Omega \subset \mathbb{R}^n$. For each cell $K \in \Omega_h$ we define

$r_K =$
$\sup[r > 0, \exists \mathbf{x} \in K : \{\mathbf{y} \in \mathbb{R}^n : |\mathbf{y} - \mathbf{x}| < r\} \subset K]$

that is, r_K stands for the radius of the largest ball that fits into K. Then the family $\{\Omega_h\}_{h \in \mathbb{H}}$ is called *shape regular* if there exists a constant $C > 0$ such that

$$\sup\{\operatorname{diam}(K)/r_K, K \in \Omega_h, h \in \mathbb{H}\} \le C.$$

This constant is sometimes referred to as shape regularity measure, or shortly, shape regularity. In classical finite element theory shape regularity in conjunction with *affine equivalence* is a crucial prerequisite for proving asymptotic estimates for approximation errors and interpolation errors of families of *finite element spaces*.

shear stress, τ   Force acting tangentially to a surface divided by the area of the surface.

Sherman-Morrison formula   The useful identity

$$[A + ab']^{-1} = A^{-1} - \frac{[A^{-1}]ab'[A^{-1}]}{1 + b'[A^{-1}]a}$$

where A is a nonsingular $n \times n$ *matrix* and a and b are n-*vectors*.

shortest path   In a graph or network, this is a path from one *node* to another whose total cost is the least among all such paths. The cost is usually the sum of the arc costs, but it could be another *function* (e.g., the product for a reliability problem, or max for a fuzzy measure of risk). There are some particular labeling *algorithms* given.

signomial   The difference between two *posynomials*. This class of function defines the general geometric program.

simplex (pl. *simplices*) $\{x \in \mathbb{R}^{n+} : \sum x_j = 1\}$. For $n = 1$, this is a point ($x = 1$). For $n = 2$, this is a line segment, joining points $(1, 0)$ and $(0, 1)$. For $n = 3$, this is a triangle, joining the vertices $(1, 0, 0)$, $(0, 1, 0)$, and $(0, 0, 1)$. This is sometimes called an *n-simplex*, denoted by S_n

(note its dimension is $n - 1$). The *open simplex* excludes the axes: $\{x \in S_n : x > 0\}$.

More generally, some authors define a *simplex* to be the convex hull of any $n + 1$ affinely independent *vectors*, and refer to the special case of the unit vectors as the *standard simplex*.

simplex method   An *algorithm* invented to solve a linear program by progressing from one extreme point of the feasible polyhedron to an adjacent one. The method is an *algorithm strategy*, where some of the tactics include *pricing* and *pivot selection*.

The *elementary simplex method* is the name of Dantzig's original (1947) algorithm, with the following rules applied to the standard form: Min $\{cx : Ax = b, x \ge 0\}$.

Let d_j = reduced cost of x_j; terminate if $d_j \ge 0$ for all j.

(i.)   Select $d_j < 0$ as one of greatest magnitude.

(ii.)   In the associated column (j) of the tableau, compute the *min ratio*: $x_i/a(i, j)$: $a(i, j) > 0$. (If $a(., j) \le 0$, LP is unbounded).

(iii.)   Enter x_j into the basic set, in exchange for x_i, and update the tableau.

Among the variations are:

(i.)   select the incoming variable (j) differently;

(ii.)   select the outgoing variable (i) differently, especially to avoid cycling; and

(iii.)   do not maintain a tableau (use a factored form of the basis).

The *revised simplex method* is the use of a particular factored form of the basis: $B = [E_1 E_2 \ldots E_k]$ (after k iterations), where each E_i is an elementary matrix. Then, the revised *simplex method* uses forward transformation to pivot and backward transformation to update the pricing vector. (For this distinction, the elementary simplex method is sometimes called the *tableau method*.)

The *simplex method* draws its name from imagining a normalization constraint, $\sum_j x_j = 1$, and thinking of the jth column of A to be selected by the weight x_j in S_w. Then, at an iteration, an m-simplex is specified by the basic variables, and an adjacent simplex is chosen to improve the objective value. This view is in requirements space.

121

simplicial subdivision   Given a simplex, S, its *simplicial subdivision* is a collection of simplices, say $\{T_i\}$ such that $\vee_i\{T_i\} = S$ and for any i and j, either the intersection $T_i \wedge T_j$ is empty or equals the closure of a common face. The *mesh* of the subdivision is the diameter of the largest subsimplex. This arises in a fixed point approach to compute an economic equilibrium.

simulated annealing   An *algorithm* for solving hard problems, notably combinatorial optimization, based on the metaphor of how annealing works: *reach a minimum energy state upon cooling a substance, but not too quickly in order to avoid reaching an undesirable final state.* As a heuristic search, it allows a non-improving move to a *neighbor* with a probability that decreases over time. The rate of this decrease is determined by the *cooling schedule*, often just a parameter used in an exponential decay (in keeping with the thermodynamic metaphor). With some (mild) assumptions about the cooling schedule, this will converge in probability to a global optimum.

sine   The function

$$\sin(x) = \frac{e^{ix} - e^{-ix}}{2i}$$

Geometrically, it is the ratio of the lengths of opposite side to hypotenuse of a right triangle with an angle x, for $0 < x < \frac{\pi}{2}$.

singleton node   A node of degree one. Synonymous with *pendant node*.

sinistralateral   The set of obligatorily coreacting species arbitrarily written on the left-hand side of a formal reaction equation.

   Comment: See *dextralateral* for explanatory comments. See also *direction, dynamic equilibrium, formal reaction equation, microscopic reversibility, product, rate constant, reversibility,* and *substrate.*

skew symmetric matrix (A) A is square and $A' = -A$.

slack variable   In an inequality constraint of the form $g(x) \leq b$, the slack is $b - g(x)$, which is designated by the slack variable, s. Then, the original constraint is equivalent to the defining equation, $g(x) + s = b$, plus $s \geq 0$.

Slater's (interiority) condition   Originally for the purely inequality system with g convex, it means there exists x for which $g(x) < 0$. More generally, for a mathematical program in standard form, it means there exists x in X for which $g(x) < 0$ and $h(x) = 0$.

slope   Of a nonvertical straight line with equation $y = mx + b$ is by definition the number m.

smetic state   See *liquid-crystal transitions* and *mesomorphic phase.*

smooth   Referring to a continuously differentiable function.

solitons   There are types of wave equations called *soliton type equations* that admit special *solitary wave solutions* called *solitons*. These *solitons* have the following property: suppose two *solitons* are moving left to right, well separated with the smaller one to the right. After some time the bigger one catches up, the waves overlap and interact. Still later the bigger wave separates from the smaller one, and eventually regains its initial shape and velocity. The only effect of the interaction is a phase shift.

   Example: The *Korteweg-deVries* equation has *solitons* of the form

$$u = 2k^2 sech^2 k(x - 4k^2 t - x_0), k, x_0 \text{ constants.}$$

spanning tree (problem)   A subgraph that is a tree containing all nodes. The *max weight spanning tree problem* is to find a *spanning tree* such that the sum of (given, positive) weights of the *edges* is a maximum.

   The max spanning tree problem is solvable by the following greedy algorithm:
   Input. Connected graph with weights, $w_1 \geq \ldots \geq w_m$.
   Output. Maximum weight spanning tree, T.

   (i.) *Initialization:* Set $k = 1$; T = graph with given nodes and no edges.

   (ii.) *Iteration* (until $k = m - 1$): Test if the kth edge forms a cycle with T. If not, add it to T; if so, discard the edge. In either case, increment k and iterate.

sparsity   The fraction of zeros in a *matrix*. If A is m by n, and $A(i, j) \neq 0$ for k of its elements, its *sparsity* is k/mn. Large linear programs tend to be very sparse, increasing as the dimensions get large. For example, consider the standard transportation problem with s sources and d destinations. This has $m = (s + d)$ constraints and $n = sd$ variables. Each column, however, has exactly two nonzeros since A is the incidence matrix of the network, so its *sparsity* is $2n/mn$, or simply $2/m$, which decreases as the number of sources and/or destinations grows large. The *sparsity* of a simple graph (or network) is the *sparsity* of its adjacency matrix. More generally, the sparsity of a multigraph refers to the average degree of its *nodes*.

specially ordered set (SOS)   Certain sets of nonnegative variables that are required to sum to 1. For computational efficiency, it is sometimes better to define these sets by some marking data structure, rather than include them along with other equality constraints. There are two types of SOSs, distinguished by what they represent. A *Type 1* SOS is one in which each variable is binary, so the constraint is one of *multiple choice*. A *Type 2* SOS is one in which a restricted basis entry rule is used, as in the lambda-form of separable programming.

specifically labeled   An *isotopically labeled* compound is designated as specifically labeled when a unique *isotopically substituted* compound is formally added to the analogous *isotopically unmodified* compound. In such a case, both position(s) and number of each labeling *nuclide* are defined.

spectral radius   (of a matrix, A) The radius of the following disk that contains the spectrum: $r(A) = \text{Max}\{|y| : y$ is an eigenvalue of $A\}$.

spectral responsivity function   See *responsivity*.

spectrum   Let T be a *linear operator* on a *Banach space* V. A complex number λ is said to be in the *resolvent set* $\rho(T)$ of T if $\lambda I - T$ is a bijection with bounded inverse. $R_\lambda(T) = (\lambda I - T)^{-1}$ is called the *resolvent* of T at λ. If $\lambda \notin \rho(T)$, then λ is said to be in the *spectrum* $\sigma(T)$ of T. The set of all *eigenvalues* of T is called the *point spectrum* of T.

spectrum of a matrix   The set of *eigenvalues* of A.

stability region   The set of parameter values for which an optimal solution remains optimal. This arises naturally in combinatorial optimization, where a solution is often a subgraph, such as a tree, and the question is for what range of arc weights is this subgraph optimal (such as a spanning tree that is minimum for given weights). More generally, x could be a solution generated by some algorithm, A, from an initial value x^0. Then, suppose the feasibility region $F(p)$ depends on the parameter p and the objective $f(x_j p)$ also depends on p. Let $X(p, A, x^0)$ denote the generated solution from *algorithm A*, starting at x^0, with parameter value p. Let the parameter set be P (which includes p^*). The *stability region* of $x^* = X(p^*, A, x^0)$ is $\{p \in P : x^* = X(p, A, x^0)\}$. The *algorithm* may be a *heuristic*, so x^* need not be optimal. For example, one could use an n-opt *heuristic* for the traveling salesman problem, so x represents a tour. The parameters could be the costs, or they could be the location of each point in a *euclidean* TSP. The stability region is the set of costs, or coordinates in the plane, for which the tour generated by n-opt is the same.

stable   As applied to *chemical species*, the term expresses a thermodynamic property, which is quantitatively measured by relative molar standard Gibbs energies. A chemical species A is more stable than its isomer B if $\Delta_r G^0 > 0$ for the (real or hypothetical) reaction $A \to B$, under standard conditions. If for the two reactions:

$$P \to X + Y \quad (\Delta_r G_1^0)$$
$$Q \to X + Z \quad (\Delta_r G_2^0)$$

$\Delta_r G_1^0 > \Delta_r G_2^0$, P is more stable relative to the product Y than is Q relative to Z. Both in qualitative and quantitative usage the term stable is therefore always used in reference to some explicitly stated or implicitly assumed standard.

The term should not be used as a synonym for *unreactive* or "less reactive" since this confuses thermodynamics and kinetics. A relatively more stable chemical species may be more *reactive* than some reference species toward a given reaction partner.

stable mathematical program   Roughly, one whose solution does not change much under perturbation. For the inequality case, we have the following stability conditions:

(i.) $\{x \in X : g(x) \le b\}$ is bounded for some $b > 0$.

(ii.) $cl\{x \in X : g(x) < 0\} = \{x \in X : g(x) \le 0\}$.

The first stability condition pertains to upper semicontinuity and the second, called the *closure condition*, pertains to lower semicontinuity.

The conditions are not only sufficient to ensure the respective semicontinuity, but they are necessary when:

(i.) $\{x \in X : g(x) \le 0\}$ is bounded,

(ii.) $\{x \in X : g(x) < 0\}$ is not empty.

standard deviation, s   The positive square root of the sum of the squares of the deviations between the observations and the mean of the series, divided by one less than the total number in the series. The standard deviation is the positive square root of the variance, a more fundamental statistical quantity.

state   A *state* of a system, $s_j \in \mathcal{S}$, where $\mathcal{S}$ is the set of states of the system, is a distinct observable or derivable variable.

Comment: The definition is meant to cover both the intensive states of thermodynamics and the states of computations and computational devices. By these definitions Post production systems have neither memory nor states; instead, the set of constructs $\mathcal{C}$ formed during a derivation is discussed.

stationary point   Usually this is used to mean a Kuhn-Tucker point, which specializes to one for which $\operatorname{grad} f(x) = 0$ if the mathematical program is unconstrained. In the context of an *algorithm*, it is a fixed point.

stationary policy   In a dynamic program, a policy that is independent of time, i.e., $x^*(s, t) = T(s)$ (some function of state, but not of time, t).

statistical genetics   Genetics is a stochastic process. Statistical genetics studies the genetics by using the concepts and methods from the theory of probability and statistics. See *population genetics*.

steady state (stationary state)   In a kinetic analysis of a complex reaction involving *unstable* intermediates in low concentration, the rate of change of each such intermediate is set equal to zero, so that the *rate equation* can be expressed as a function of the concentrations of *chemical species* present in macroscopic amounts. For example, assume that X is an unstable intermediate in the reaction sequence:

$$A + B A \underset{k_{-1}}{\overset{k_1}{\rightleftarrows}} X$$

$$X + C \overset{k_2}{\longrightarrow} D.$$

Conservation of mass requires that:

$$[A] + [X] + [D] = [A]_0$$

which, since $[A]_0$ is constant, implies:

$$-d[X]/dt = d[A]/dt + d[D]/dt.$$

Since $[X]$ is negligibly small, the rate of formation of D is essentially equal to the rate of disappearance of A, and the rate of change of $[X]$ can be set equal to zero. Applying the steady state approximation ($d[X]/dt = 0$) allows the elimination of $[X]$ from the kinetic equations, whereupon the rate of reaction is expressed:

$$d[D]/dt = -d[A]/dt = \frac{k_1 k_2 [A][C]}{k_{-1} + k_2 [C]}$$

Notes: (**1**) The *steady state* approximation does not imply that $[X]$ is even approximately constant, only that its absolute rate of change is very much smaller than that of $[A]$ and $[D]$. Since according to the reaction scheme $d[D]/dt = k_2[X][C]$, the asumption that $[X]$ is constant would lead, for the case in which C is in large excess, to the absurd conclusion that formation of the product D will continue at a constant rate even after the reactant A has been consumed.

(**2**) In a stirred flow reactor a *steady state* implies a regime in which all concentrations are independent of time.

steel beam assortment problem   A steel corporation manufactures structured beams of a standard length, but a variety of strengths. There is a known demand of each type of strength, but a stronger one may fulfill demand (or part thereof) for another beam (but not conversely). The manufacture of each type of steel beam involves a fixed charge for its setup. In addition, there is a shipping cost proportional to the difference in the demanded strength and the actual strength, and proportional to the quantity shipped.

Let

$$
\begin{aligned}
N &= \text{number of varieties of strengths} \\
D(t) &= \text{demand for beam of strength } s_t \\
&\quad (\text{where } s_1 \geq s_2 \geq \ldots \geq s_N) \\
x(t) &= \text{amount of beams of strength } s_t \\
&\quad \text{manufactured} \\
p_t(x(t)) &= \text{manufacturing cost of x(t) units} \\
&\quad \text{of beam of strength } s_t \text{ (including)} \\
&\quad \text{fixed charge} \\
y(t) &= \text{total excess of beams of strength} \\
&\quad s_1, \ldots, s_t \text{ before fulfilling demand} \\
&\quad D(t+1), \ldots, D(N) \\
h_t(y(t)) &= \text{shipping cost}(= c[s_(t+1) - s_t] \\
&\quad \text{Min}\{y(t), D(t)\}).
\end{aligned}
$$

Although t does not index time, the mathematical program for this problem is of the same form as the production scheduling problem, using the inventory balance equations to relate y and x. This is valid because $s_1 \geq s_2 \geq \ldots \geq s_N$ implies $y(t)$ can be used to fulfill demand $D(t+1) + D(t+2) + \ldots + D(N)$. (Also, note that here h_t is not a holding cost.)

steepest ascent (descent, if minimizing)   A class of *algorithms*, where $x' = x + sd$, such that the direction vector d is chosen by maximizing the initial velocity of change, and the step size (s) is chosen by line search. Generally used in the context of unconstrained optimization, the mathematical program is Max$\{f(x) : x \in \mathbb{R}^n\}$, where f is in C^1. (For *descent*, change Max to Min.) Then, d is chosen to maximize the first-order Taylor approximation, subject to a normalization constraint: Max$\{\text{grad} f(x)d : \|d\| = 1\}$, where $\|d\|$ denotes the norm of the direction vector, d. When the Euclidean norm is used, this yields the original steepest ascent algorithm by Cauchy, which moves in the direction of the gradient:

$$
d = \text{grad} f(x)/\|\text{grad} f(x)\|.
$$

(No direction vector is sought if grad $f(x) = 0$; such algorithms stop when reaching a stationary point.)

Other norms, such as $\|d\|^2 = d' Q d$, where Q is symmetric and positive definite, lead to other directions that are *steepest* relative to that norm. In particular, if $Q = H_f(x)$, this yields the modified Newton method.

Steiner problem   Find a subgraph of a graph, say $G' = [V', E']$, such that V' contains V^* (a specified subset of nodes), and Sum$\{c(e) : e \in E'\}$ is minimized. It is generally assumed $c \geq 0$. When $|V^*| = 2$, this is the shortest path problem. When $|V^*| = |V|$, this is the (minimum) spanning tree problem.

step size   A scalar (s) in an *algorithm* of the form: $x' = x + sd$, where d is the *direction vector*. After d is chosen (nonzero), the step size is specified. One step size selection rule is line search; another is a fixed sequence, like $s_k = 1/k$.

stepwise reaction   A *chemical reaction* with at least one reaction *intermediate* and involving at least two consecutive *elementary reactions*.

stereochemical formula (stereoformula)   A three-dimensional view of a molecule either as such or in a projection.

sticking coefficient   (in surface chemistry) The ratio of the rate of *adsorption* to the rate at which the *adsorptive* strikes the total surface, i.e., covered and uncovered. It is usually a function of surface coverage, of temperature and of the details of the surface structure of the *adsorbent*.

stiffness matrix   The *Galerkin discretization* of a linear variational problem by means of a *finite element space* V_h leads to a linear system of equations. If a stands for the *sesqui-linear* form of the variational problem, the system matrix is given by

$$
A := (a(b_i, b_j))_{i,j=1}^N,
$$

where $\{b_1, \cdots, b_N\}$, $N = \dim V_h$, is the *nodal basis* of V_h. If the *sesqui-linear* form a arises from the *weak formulation* of a boundary value problem for a partial differential equation, the interaction of the basis functions will be local in the sense that

$$\text{meas}(\text{supp } b_i \cap \text{supp } b_j) = 0 \Rightarrow a(b_i, b_j) = 0.$$

This leads to considerable *sparsity* of the stiffness matrix, which is key to the efficiency of finite element schemes.

stochastic computation   Let the set of symbols of the computation be Σ, and the probability that a particular symbol σ_i exists be $P_e(\sigma_i)$. Then a computation C is executed *stochastically* if for at least one $\sigma_i \in \Sigma$, $P_e(\sigma_i) \leq 1$; or if at least one $c_i \in C$ is a stochastic function.

Comment: This definition explicitly posits that determinism is a property of a computation and stochasticity that of its execution. An example of the latter is when at least one σ_i has a half-life which is not orders of magnitude greater than that of t_C, the time needed to execute the computation, so that one cannot rely upon σ_i to persist throughout the lifetime of the computation (including the step of actually detecting it). See *deterministic* and *nondeterministic computation*.

stochastic matrix   A nonnegative matrix whose row sums are each 1. (A *column stochastic matrix* is one whose column sums are 1.) This arises in dynamic programs whose state transition is described by a *stochastic matrix* containing the probabilities of each transition.

stochastic program   A program written in the form of a mathematical program extended to a parameter space whose values are random variables (generally with a known distribution function). This is converted to a standard form by forming a *certainty equivalent*. Here are some certainty equivalents:

(**1**) *Average value.* Replace all random variables with their means.

(**2**) *Chance constraint.* Given a *stochastic program* with a random variable, p, in its constraint: $g(x; p) \leq 0$, a *certainty equivalent* is to replace this with the constraint, $P[g(x; p) \leq 0] \geq a$, where $P[\ ]$ is a (known) probability function on the range of g, and a is some *acceptance level* ($a = 1$ means the constraint must

hold for *all* values of p, except on a set of measure zero). Some models separate constraints with several levels:

$$P[g_i(x; p) \leq 0 \text{ for all } i \text{ in } I_k] \geq a_k$$

$$\text{for } k = 1, ..., K.$$

The case of one chance constraint with the only random variable being the right-hand side is particularly simple. Suppose F is the cumulative distribution function of b for the chance constraint $P[g(x) \leq b] \geq a$. If b is a continuous random variable and F is continuous and strictly increasing, the chance constraint is equivalent to $g(x) \leq F^{-1}(1 - a)$ (where F^{-1} is the inverse function of F). In particular, if $g(x) = Ax$, the program remains linear.

(**3**) *Recourse model.* This assumes decisions are made over time where the effect of an early decision can be compensated by later decisions. The objective is to optimize the expected value. The two-stage model has the form

$$\text{Max } f_1(x_1; p_1) + f_2(x_2; p_2) :$$

$$x_1 \in X_1, x_2 \in X_2, g(x_1; p_1) + g(x_2; p_2) \leq 0.$$

(Sums could be replaced by other operators.) Once x_1 is implemented, p_1 becomes known and x_2 is chosen. The certainty equivalent is

$$\text{Max } E[f_1(x_1; p_1) + F_2(x_1|p_1)] : x_1 \in X_1,$$
where

$$F_2(x_1|p_1) = \text{Sup}\{E[f_2(x_2; p_2)] :$$

$$x_2 \in X_2(p_2), g(x_2; p_2) \leq -g(x_1; p_1)\}$$

for all p_2 (except on set of measure zero). ($E[\ ]$ denotes the expected value.) The "Sup" is used to define F_2, the second stage value for a particular value of x_1, because the choice of x_1 might be infeasible. The nature of the recourse model is pessimistic: x must be chosen such that the original constraints hold no matter what the values of the random variables. With a finite number of possibilities, this means a system of constraints for each possible realization of $p = (p_1, p_2)$. This sextends recursively to a k-stage model.

The linear two-stage recourse model has the form:

$$\text{max } E[c]x + E[F(x; p)] :$$

$$Ax = b, x \geq 0,$$

where

$$F(x; p) = \max d(p)y : W(p)y$$

$$= w(p) - T(p)x, y \geq 0.$$

Here the second stage variable is denoted y. It is determined after x has been set and the random variable p has been realized. The LP data depend on p as functions, $d(p)$, $W(p)$, $w(p)$, and $T(p)$. The *fixed recourse model* has $W(p) = W$. The *complete recourse model* assumes it is fixed and $\{Wy : y \geq 0\}$ is all of $\mathbb{R}^m$ (where $m =$ number of rows of W). This means that no matter what value of x is chosen for the first stage, there is feasible recourse (y). This is *simple recourse* if $W = [I - I]$, so we can think of y as having two parts: y^{pos} and y^{neg}. The second stage LP simplifies to the following:

$$\max d^{\text{pos}}(p)y^{\text{pos}} + d^{\text{neg}}(p)y^{\text{neg}} :$$

$$y^{\text{pos}}, y^{\text{neg}} \geq 0$$

$$y^{\text{pos}} - y^{\text{neg}} = w(p) - T(p)x.$$

The certainty equivalent depends upon the underlying decision process. If it is adaptive, the recourse model applies (but RO might be more practical). The chance constraint model represents a notion of an allowed frequency of violations, as in environmental control models.

stoichiometry   The number of moles of a reactant used in a reaction, normalized to the number of *moles* of all the other reactants.

Comment: Stoichiometries are always integers, because they are determined by chemical indivisibility of atoms. See *mole*.

Strang's lemmas   Consider a linear variational problem $a(u, v) = f(v), v \in V$ posed over a Hilbert space V. Commiting a *variational crime* the corresponding discrete variational problem reads

$$u_h \in V_h : a_h(u_h, v_h) = f_h(v_h) \ \forall v_h \in V_h,$$

where $V_h \subset V$ is a V-conforming finite element space, $a_h : V_h \times V_h \to \mathbb{C}$ a *sesqui-linear* form, and $f_h : V_h \to \mathbb{C}$ a linear form. If a_h is *V-elliptic*, that is,

$$|\mathcal{R}\{a_h(v_h, u_h)\}| \geq \alpha \|v_h\|_V^2 \ \forall v_h \in V_h,$$

then the first of *Strang's lemmas* tell us

$$\|u - u_h\|_V$$

$$\leq C(\inf_{v_h \in V_h}(\|u - v_h\|_V$$

$$+ \sup_{w_h \in V_h} \frac{|a(v_h, w_h) - a_h(v_h, w_h)|}{\|w_h\|_V})$$

$$+ \sup_{w_h \in V_h} \frac{|f(w_h) - f_h(w_h)|}{\|w_h\|_V}),$$

with $C = C(\alpha, \|a\|, \|a_h\|) > 0$. The second *Strang's lemma* targets a *non-conforming* choice of V_h, that is $V_h \not\subset V$. As $\|.\|_V$ is not necessarily well defined for functions of V, we have to introduce a *mesh-dependent norm* $\|.\|_h$ on $V_h + V$ for which a_h is *elliptic*

$$|\mathcal{R}\{a_h(v_h, u_h)\}| \geq \alpha \|v_h\|_h \ \forall v_h \in V_h.$$

Moreover, we have to require continuity

$$|a_h(u, v_h)| \leq c\|u\|_h\|v_h\|_h$$

$$\forall u \in V_h + V, \ v_h \in V_h.$$

Then with $C = C(\alpha, c) > 0$

$$\|u - u_h\|_V \leq C(\inf_{v_h \in V_h}\|u - v_h\|_V$$

$$+ \sup_{w_h \in V_h} \frac{|a(v_h, w_h) - a_h(v_h, w_h)|}{\|w_h\|_V}).$$

stratopause   That region of the atmosphere which lies between the *stratosphere* and the *mesosphere* and in which a maximum in the temperature occurs.

stratosphere   The atmospheric shell lying just above the *troposphere* which is characterized by an increasing temperature with altitude. The *stratosphere* begins at the *tropopause* (about 10–15 km height) and extends to a height of about 50 km, where the lapse rate changes sign at the stratopause and the beginning of the *mesosphere*.

strict interior   Let $\{x \in X : g(x) \leq b\}$ be the level set of g. Then, its *strict interior* is $\{x \in X : g(x) < b\}$. (This is not to be confused with the relative interior or the set interior. See *interior*.)

strictly complementary   Each complementary pair of variables must have exactly one zero (the other positive).

strictly concave function   Negative of a *strictly convex function*.

strictly convex function   A convex function that also satisfies the defining inequality strictly for distinct points, say x and y:

$$f(ax + (1 - a)y) < af(x) + (1 - a)f(y)$$

for all a in $(0, 1)$.

strictly quasiconcave function   Negative of a *strictly quasiconvex function*.

strictly quasiconvex function X is a convex set and $f(ax + (1 - a)y) < \text{Max}\{f(x), f(y)\}$ for all x, y in X for which $f(x) \neq f(y)$, and a is in $(0, 1)$. *Note:* f need not be quasiconvex.

strong collision   A collision between two molecules in which the amount of energy transferred from one to the other is large compared with $k_B T$, where k_B is the Boltzmann constant and T the absolute temperature.

strongly concave function   Negative of a strongly convex function.

strongly convex function   A function f in C^2 with eigenvalues of its Hessian bounded away from zero (from below): there exists $K > 0$ such that $h' H_f(x)h \geq K\|h\|^2$ for all h in $\mathbb{R}^n$. For example, the function $1 - \exp(-x)$ is strictly convex on $\mathbb{R}$, but its second derivative is $-\exp(-x)$, which is not bounded away from zero. The minimum is not achieved because the function approaches its infimum of zero without achieving it for any (finite) x. Strong convexity rules out such asymptotes.

strongly quasiconcave function   Negative of a *strongly quasiconvex function*.

strongly quasiconvex function   (f on X) On a convex set X $f(ax + (1 - a)y) < \text{Max}\{f(x), f(y)\}$ for all x, y in X, with $x \neq y$, and a in $(0, 1)$.

subadditive function $f(x + y) \leq f(x) + f(y)$ where x, y in the domain implies $x + y$ is in the domain.

subbag   See *parts of collections*.

subdifferential (of f at x) $\partial_f(x) = \{y : x$ is in argmax $\{v_y - f(v) : v \in X\}\}$. If f is convex and differentiable with gradient, grad f, $\partial_f(x) = \{\text{grad } f(x)\}$. Example: $f(x) = |x|$. Then, $\partial_f(0) = [-1, 1]$.

The subdifferential is built on the concept of supporting hyperplane, generally used in convex analysis. When f is differentiable in a deleted *neighborhood* of x (but not necessarily at x), the *B-subdifferential* is the set of limit points:

$$\partial_B f(x) = \{d : \text{ there exists } \{x^k\} > x$$

$$\text{and } \{\text{grad } f(x^k)\} > d\}.$$

If f is continuously differentiable in a *neighborhood* of x (including x), $\partial_B f(x) = \{\text{grad } f(x)\}$. Otherwise, $\partial_B f(x)$ is generally not a convex set. For example, if $f(x) = |x|$, $\partial_B f(0) = \{-1, 1\}$.

The *Clarke subdifferential* is the convex hull of $\partial_B f(x)$.

subgradient   A member of the subdifferential.

subgraph   A graph $G'(\mathcal{V}', \mathcal{E}')$ is a *subgraph* of $G(\mathcal{V}, \mathcal{E})$ if every *node* and *edge* present in G' is present in G; that is, $\mathcal{V}' \subseteq \mathcal{V}$ and $\mathcal{E}' \subseteq \mathcal{E}$. G' is a *proper subgraph* of G if $G' \neq G$.

sublist   See *parts of collections*.

submodular function   Let N be a finite set and let f be a function on the subsets of N into $\mathbb{R}$. Then, f is *submodular* if it satisfies:

$$f(S \wedge T) \leq f(S) + f(T) - f(S \wedge T)$$

for S, T subsets of N.

subnetwork   A network $N'(\mathcal{V}', \mathcal{E}', \mathcal{P}', \mathcal{L}')$ is a *subnetwork* of $N(\mathcal{V}, \mathcal{E}, \mathcal{P}, \mathcal{L})$ if every node, edge, parameter, and label present in N' is present in N; that is, $\mathcal{V}' \subseteq \mathcal{V}$; $\mathcal{E}' \subseteq \mathcal{E}$; $\mathcal{P}' \subseteq \mathcal{P}$; and $\mathcal{L}' \subseteq \mathcal{L}$. N' is a *proper subnetwork* of N if also $N' \neq N$.

subsequence   A subset of a *sequence*, with the order preserved.

subset   See *parts of collections.*

subspace   A subset of a *vector space* that is, itself, a *vector space.* An example is the null space of a *matrix*, as well as its orthogonal complement.

substituent atom (group)   An atom (group) that replaces one or more hydrogen atoms attached to a parent structure or characteristic group except for hydrogen atoms attached to a chalcogen atom.

substitution   In logic and logic programming, a substitution θ is a finite set of pairs of the form $X_i = t_i$, where $X_i \in X$ are unique variables ($X_i \neq X_j$ for all $i \neq j$ and $X_i \notin t_j$ for any i and j), and $t_i \in t$ are ground terms. Thus, the result, A', of applying a substitution θ to a complex term A, denoted $A\theta$, is the term obtained by replacing each X_i in A by t_i in the new term A', for every pair $X_i = t_i \in \theta$.
    Comment: It is important to realize that θ is a set of mappings or transformations (substitutions) between the variables in the variable-containing term in the original language, and a term from the transformed language with all variables fully substituted. This mapping produces the ground terms t. Thus, one can as well write $\theta = X_1 = t_1, X_2 = t_2, \ldots, X_n = t_n$. Do not confuse this usage with the chemical meaning of substitution.

substitution reaction   A *reaction, elementary* or *stepwise*, in which one atom or group in a *molecular entity* is replaced by another atom or group. For example,

$$CH_3Cl + OH^- \rightarrow CH_3OH | Cl^-$$

substrate   (**1**) (in biochemistry) The specific molecules which are recognized by an enzyme.
    (**2**) (in chemistry) A *chemical species*, the reaction of which with some other chemical reagent is under observation (e.g., a compound that is transformed under the influence of a *catalyst*).
    Comment: The term should be used with care. Either the context or a specific statement should always make it clear which chemical species in a reaction is regarded as the substrate. The distinction is between the molecular material of a

reaction and where a compound is represented in the arbitrarily written formal reaction equation. See also *dextralateral, direction, dynamic equilibrium, formal reaction equation, microscopic reversibility, product, rate constant, reversibility*, and *sinistralateral.*

successive approximation   The iterative scheme by which an approximation is used for the basic design of an *algorithm.* The sequence generated is of the form $x^{(k+1)} = x^k + A(x^k)$, where A is an *algorithm* map specified by its approximation to some underlying goal. Typically, this is used to find a fixed point, where $A(x) = 0$ (e.g., seeking $f(x) = x$, let $A(x) = f(x) - x$, so the iterations are $x^{(k+1)} = f(x^k)$, converging to $x^* = f(x^*)$ if f satisfies certain conditions, such as being a contraction map).

Here are some special types:

(i.)   Inner approximation

(ii.)   Outer approximation

(iii.)   Successive linear approximation

(iv.)   Successive quadratic approximation

sufficient matrix   Let A be an $n \times n$ matrix. Then, A is *column sufficient* if

$$[x_i(A_x)_i \leq 0 \text{ for all } i] \Rightarrow$$

$$[x_i(A_x)_i = 0 \text{ for all } i].$$

A is *row sufficient* if its *transpose* is column sufficient. A is sufficient if it is both column and row sufficient. One example is when A is symmetric and positive semidefinite. This arises in linear complementarity problems.

superset   A set which contains another set. If the *superset* and the contained set can be equal, the relation between them is denoted superset $\supseteq$ contained set; otherwise it is denoted $\supset$.
    Comment: The same idea can be applied to other types of collections, just considering *parts of collections* in the opposite sense. See also *bag, empty collection, list, parts of collections, sequence*, and *set.*

surface tension, γ, σ   Work required to increase a surface area divided by that area. When two phases are studied it is often called interfacial tension.

surjection   A map $\phi : A \to B$ which is *surjective*.

surjective   Let A and B be two sets, with A being the domain and B the codomain of a function f. Then the function f is *surjective* if, for any $y \in B$, there is *at least* one element $x \in A$ with $f(x) = y$. Thus, the range of a surjective functions is equal to its codomain. Surjective functions are also said to be *onto B*, and may be many-to-one. See also *bijection* and *injection*.

symmetric difference   Given two sets, A and B, their *symmetric difference* is defined as $A \otimes B = (A - B) \cup (B - A)$.

Comment: To visualize this more clearly, think about the elements of the two difference sets, $A - B$ and $B - A$. An element $d_1 \in A - B$ if and only if $d_1 \in A$ and $d_1 \notin B$. This must be true for all of the elements of $A - B$, by the definition for set difference. Similarly, for an element $d_2 \in B - A$. So the intersection of the two difference sets, $A - B$ and $B - A$ must be empty: any elements shared between A and B would already be removed by the set difference operation. So what is left over are the elements which are present in only one of the two sets, for each of the two sets. See also *set difference*.

symmetric operator   An operator T, densely defined on a *Hilbert space*, satisfying $< Tu, v > = < u, Tv >$ for all u, v in the domain of T.

symplectic manifold   A pair (P, ω) where P is a *manifold* and ω is a closed non-degenerate 2-form on P. As a consequence P is of even dimension $2n$.

Canonical coordinates are coordinates (q^{λ}, p_{λ}) such that the local expression of ω is of the form $\omega = \mathrm{d}p_{\lambda} \wedge \mathrm{d}q^{\lambda}$. Canonical coordinates always exist on a symplectic manifold (Darboux theorem).

syntax   For a symbol σ, a unique and precise definition of the form of the term and all terms composing it.

Comment: Syntax defines such things as whether a term is a constant or a variable, how many arguments it has, and the syntax of those arguments.

T

tableau (pl. *tableaux*)   A detached coefficient form of a system of equations, which can change from $x + Ay = b$ to $x' + A'y' = b'$. The primes denote changes caused by multiplying the first equation system by the basis inverse (a sequence of pivots in the simplex method). Other information could be appended, such as the original bound values.

tabu search   This is a metaheuristic to solve global optimization problems, notably combinatorial optimization, based on multilevel memory management and response exploration. It requires the concept of a *neighborhood* for a trial solution (perhaps partial). In its simplest form, a *tabu search* appears as follows:

(i.) Initialize. Select x and set *Tabu List* $T =$ null. If x is feasible, set $x^* = x$ and $f^* = f(x^*)$; otherwise, set $f^* = -\inf$ (for minimization set $f^* = \inf$).

(ii.) Select move. Let $S(x) =$ set of *neighbors* of x. If $S(x)\backslash T$ is empty, go to update. Otherwise, select y in argmax $\{E(v) : v$ in $S(x)\backslash T\}$, where E is an evaluator function that measures the merit of a point (need not be the original objective function, f). If y is feasible and $f(y) > f^*$, set $x^* = y$ and $f^* = f(x^*)$. Set $x = y$ (i.e., move to the new point).

(iii.) Update. If some stopping rule holds, stop. Otherwise, update T (by some tabu update rule) and return to select move.

There are many variations, such as aspiration levels, that can be included in more complex specifications.

tangent (function)   The function

$$\tan(x) = \frac{\sin x}{\cos x}.$$

See *cosine, sine*.

tangent cone   Let S be a subset of $\mathbb{R}^n$ and let x^* be in S. The *tangent* cone, $T(S, x^*)$, is the set of points y such that there exist sequences $\{a_n\}, a_n \geq 0$ and $\{x^n\}$ in S such that $\{x^n\} \to x^*$ and $\{\|a_n(x^n - x^*) - y\|\} \to 0$. This arises in connection with the Lagrange multiplier rule much like the *tangent plane*, though it allows for more general constraints, e.g., set constraints. In particular, when there are only equality constraints, $h(x) = 0$, $T(S, x^*) =$ null space of grad $h(x^*)$ if grad $h(x^*)$ has full row rank. (There are some subtleties that render the *tangent cone* more general, in some sense, than the *tangent plane* or *null space*. It is used in establishing a necessary constraint qualification.)

tangent lift   A general procedure to associate canonically an object on the tangent bundle TM of a *manifold M* once an object is given on M. In particular:

(i.) *Tangent lift of a parametrized curve γ* : $I \subset \mathbb{R} \to M$: let us define $\tau_\gamma(t_0) = \frac{d\gamma}{dt}|_{t=t_0}$ the tangent vector to the curve γ at $t = t_0 \in I$. We can define the tangent lift of γ as the curve $\hat{\gamma} : I \to TM : t \mapsto (\gamma(t), \tau_\gamma(t))$. If $\gamma^\mu(t)$ is the local expression of γ, then $(\gamma^\mu(t), \dot{\gamma}(t))$ is the local expression of $\hat{\gamma}$.

(ii.) *Tangent lift of a map $\phi : M \to N$*: if tangent vectors are identified with derivations of the *algebra* of local functions, the tangent map $T\phi : TM \to TN : v \mapsto w$ is defined by $w(f) = v(f \circ \phi)$. If $x'^\mu = \phi^\mu(x)$ is the local expression of ϕ, then the local expression of the tangent map $T\phi$ is given by:

$$x'^\mu = \phi^\mu(x)$$
$$w'^\mu = v^\nu \partial_\nu \phi^\mu(x) K$$

(iii.) *Tangent lift of a vector field $\xi = \xi^\mu(x) \partial_\mu \in \mathfrak{X}(M)$*: a vector field $\hat{\xi} = \xi^\mu(x) \frac{\partial}{\partial x^\mu} + \hat{\xi}^\mu(x) \frac{\partial}{\partial v^\mu}$ over TM locally given by $\hat{\xi}^\mu = \xi^\nu \partial_\nu \xi^\mu$. Notice that the *tangent lift* of a commutator coincides with the commutator of lifts, i.e., $[\xi, \zeta]\hat{} = [\hat{\xi}, \hat{\zeta}]$.

tangent plane   Consider the surface, $S = \{x \in \mathbb{R}^n : h(x) = 0\}$, where h is in C^1. A differentiable curve passing through x^* in S is $\{x(t) : x(0) = x^*$ and $h(x(t)) = 0$ for all t in $(-e, e)\}$, for which the derivative, $x'(t)$, exists, where $e > 0$. The *tangent plane* at x^* is the set of all initial derivatives: $\{x'(0)\}$. (This is a misnomer, except in the special case of one function and two variables at a nonstationary point.) An important fact that underlies the classical Lagrange multiplier theorem when the rank of $\operatorname{grad} h(x^*)$ is full row (x^* is then called a *regular point*): the tangent plane is $\{d : \operatorname{grad} h(x^*) d = 0\}$.

Extending this to allow inequalities, the equivalent of the tangent plane for a regular point (x^*) is the set of directions that satisfy first-order conditions to be feasible:

$$\{d : \operatorname{grad} h(x^*)d = 0 \text{ and}$$

$$\operatorname{grad} g_i(x^*)d \le 0 \text{ for all } i : g_i(x^*) = 0\}.$$

target analysis   This is a metaheuristic to solve global optimization problems, notably combinatorial optimization, using a learning mechanism. In particular, consider a branch and bound strategy with multiple criteria for branch selection. After solving training problems, hindsight is used to eliminate dead paths on the search tree by changing the weights on the criteria: set $w > 0$ such that $wV_i \le 0$ at node i with value V_i, that begins a dead path, and $wV_i > 0$ at each node, i, on the path to the solution. If such weights exist, they define a separating hyperplane for the test problems. If such weights do not exist, problems are partitioned into classes, using a form of feature analysis, such that each class has such weights for those test problems in the class. After training is complete, and a new problem arrives, it is first classified, then those weights are used in the branch selection.

Taylor expansion   For f in C^n, Taylor's theorem is used by dropping the remainder term. The first-order expansion is $f(x) = f(y) + \operatorname{grad} f(x)(x - y)$, and the second-order expansion is $f(x) = f(y) + \operatorname{grad} f(x)(x - y) + (x - y)^t H_f(x)(x - y)/2$.

Taylor series   For a function, f, having all order derivatives, the series

$$\sum_{k=0}^{\infty} \frac{f^{(k)}(h)}{k!}(x - h)^k,$$

where $f^{(k)}$ is the kth derivative of f. Truncating the series at the nth term, the error is given by:

$$|\mathcal{E}_n(h)| = \left| f(x) - \sum_{k=0}^{n} \frac{f^{(k)}(h)}{k!}(x - h)^k \right|.$$

This is a Taylor expansion, and for the Taylor series to equal the functional value, it is necessary that the error term approaches zero for each n:

$$\lim_{h \to 0} \mathcal{E}_n(h) = 0.$$

In any case, there exists y in the line segment $[x, x + h]$ such that

$$\mathcal{E}_n(h) = \frac{f^{(n+1)}(y)}{(n + 1)!}(y - h)^{n+1}.$$

Taylor theorem   Let $f : (a - h, a + h) \to \mathbb{R}$ be in C^{n+1}. Then, for x in $(a, a + h)$,

$$f(x) = f(a) + [f^{(1)}(a)][x - a] +$$

$$\ldots + [f^{(n)}(a)][(x - a)^n]/n! + R_n(x, a),$$

where $R_n(x, a)$, called the remainder, is given by the integral:

$$\int_a^x \frac{(x - t)^n}{n!} f^{(n+1)}(t)\, dt.$$

This extends to multivariate functions and is a cornerstone theorem in nonlinear programming. Unfortunately, it is often misapplied as an approximation by dropping the remainder, assuming that it goes to zero as $x \to a$.

telegraph equation   Let $U \subset \mathbb{R}^n$ be open and $u : U \times \mathbb{R} \to \mathbb{R}$. The *telegraph equation* for u is

$$u_{tt} + du_t - u_{xx} = 0.$$

temperature inversion (in atmospheric chemistry)   A departure from the normal decrease of temperature with increasing altitude. A temperature inversion may be produced, for example, by the movement of a warm air mass over a cool one. Intense surface inversions may form over the land during nights with clear skies and low winds due to the radiative loss of heat from the surface of the earth. The temperature increases as a function of height in this case. Poor mixing of the pollutants generally occurs below the inversion, since the normal convective process which drives the warmer and lighter air at ground-level to higher altitudes is interrupted as the rising air parcels encounter the warmer air above. Temperature inversions near the surface are particularly effective in trapping ground-level emissions.

temperature jump   A relaxation technique in which the temperature of a chemical system is suddenly raised. The system then relaxes to a new state of equilibrium, and analysis of the relaxation processes provides rate constants.

temperature lapse rate (in atmospheric chemistry)   The rate of change of temperature with altitude (dT/dz). The rate of temperature decrease with increase in altitude which is expected to occur in an unperturbed dry air mass is $9.8 \times 10^3 \, °C \, min^{-1}$. This is called the dry *adiabatic lapse rate*. The lapse rate is taken as positive when temperature decreases with increasing height. For air saturated with H_2O the lapse rate is less because of the release of the latent heat of water as it condenses. The average tropospheric lapse rate is about $6.5 \times 10^3 \, °C \, min^{-1}$. The lapse rate has a negative value within an inversion layer.

tensor   See *contravariant tensor*.

term   (**1**) A variable, constant, or complex expression of the form $f(\sigma_1, \sigma_2, \ldots, \sigma_n)$ where there exists at least one σ_i (that is, the arity of f is always positive). If f is a relation (or predicate) symbol, then $f(\sigma_1, \sigma_2, \ldots, \sigma_n)$ is an *atomic formula*.

Comment: Note that this definition is recursive (a term is either a term or a function of terms), and that it includes both variable and constant (or ground) terms. It is taken to be synonymous with *token*.

(**2**) (in x-ray spectroscopy) A set of levels which have the same electron configuration and the same value of the quantum numbers for total spin S and total orbital angular momentum, L.

term semantics   If for each $\sigma_j \in \Sigma$ there is at least one semantic mapping ω that is semantically well formed, then we say the *term semantics* of $\Omega \colon \Sigma \longmapsto \Omega(\Sigma)$ are well-formed, where Ω is

the set of all semantically well-formed mappings operating on Σ and $\Omega(\Sigma)$ is the set of defined semantics for the members of Σ.

Comment: This notion of term semantics extends in the domain direction the notions of semantics of programs, and is consistent with it. See *semantics* and *semiote*.

term, T   Energy divided by the product of the Planck constant and the speed of light, when of wave number dimension, or energy divided by the Planck constant, when of frequency dimension.

terminal nodes   The first and last *nodes* in the sequence of *nodes* and *edges* forming a path. If the path is a connected tree, the last nodes are the leaves of the tree (letting the root of the tree be the first *node* of the sequence). See also *path* and *sequence*.

test sample   The sample, prepared from the *laboratory sample*, from which *test portions* are removed for testing or for analysis.

theorem of the alternative   Any of several theorems that establish that two systems are alternatives. See, for example, *Fredholm alternative*.

thermal conductance, *G*   Heat *flow rate* divided by the temperature difference.

thermal conductivity, λ   Tensor quantity relating the heat flux, J_q to the temperature gradient, $J_q = -\lambda \, \mathbf{grad} \, T$.

thermal resistance, *R*   Reciprocal of the *thermal conductance*.

thermodilatometry   A technique in which a dimension of a substance under negligible load is measured as a function of temperature while the substance is subjected to a controlled temperature program.

Linear thermodilatometry and volume thermodilatometry are distinguished on the basis of the dimensions measured.

thermodynamic isotope effect   The effect of isotopic or substitution on an equilibrium constant is referred to as a thermodynamic (or equilibrium) isotope effect. For example, the effect of isotopic substitution in reactant A that participates in the equilibrium:

$$A + B \rightleftharpoons C$$

is the ratio K^1/K^h of the equilibrium constant for the reaction in which A contains the light isotope to that in which it contains the heavy isotope. The ratio can be expressed as the equilibrium constant for the isotopic exchange reaction

$$A^1 + C^h \rightleftharpoons A^h + C^1$$

in which reactants such as B that are not isotopically substituted do not appear.

The potential energy surfaces of isotopic molecules are identical to a high degree of approximation, so thermodynamic isotope effects can only arise from the effect of isotopic mass on the nuclear motions of the reactants and products, and can be expressed quantitatively in terms of partition function ratios and for nuclear motion:

$$\frac{K^1}{K^h} = \frac{(Q^1_{nuc}/Q^h_{nuc})_C}{(Q^1_{nuc}/Q^h_{nuc})_A}.$$

Although the nuclear partition function is a product of the translational, rotational, and vibrational partition functions, the isotope effect is determined almost entirely by the last named, specifically by vibrational modes involving motion of isotopically different atoms. In the case of light atoms (i.e., protium vs. deuterium or tritium) at moderate temperatures, the isotope effect is dominated by zero-point energy differences.

thermodynamic motif   A conserved pattern of changes in the thermodynamic quantities G, H, or S for a set of reactions.   See also *biochemical, chemical, dynamical, functional, kinetic, mechanistic, phylogenetic, regulatory,* and *topological motives.*

thermolysis   The uncatalyzed cleavage of one or more covalent *bonds* resulting from exposure of a compound to a raised temperature, or a process in which such cleavage is an essential part.

thermosphere   Atmospheric shell extending from the top sof the *mesosphere* to outer space. It is a region of more or less steadily increasing temperature with height, starting at 70 or 80 km. It includes the exosphere and most or all of the ionosphere (not the D region).

threshold energy, E_0   Potential energy gap between *reactants* and the *transition state*, sometimes involving the zero point energies, but usually not.

threshold phenomenon   For a linearly stable fixed point in a system of ordinary differential equations, returning to the fixed point is monotonic for small perturbations. But for perturbations greater than a threshold, the dynamic variables can undergo large excursion before returning to the fixed point (see *excitability*).

tight constraint   Same as active constraint, but some authors exclude the redundant case, where an inequality constraint happens to hold with equality, but it is not binding.

time constant (of a detector), τ_c   If the output of a *detector* changes exponentially with time, the time required for it to change from its initial value by the fraction $[1 - \exp(-t/\tau_c)]$ (for $t = \tau_c$) of the final value, is called the time constant.

time-staged   A model with a discrete time parameter, $t = 1, ..., T$, as in dynamic programming, but the solution technique need not use the DP recursion. The number of time periods (T) is called the planning horizon.

tint   The edge coloring corresponding to the type of biochemical relationship between two nodes of the biochemical network.

Comment: The three fundamental relationships are *sinistralateral, dextralateral,* and *catalyst.* The sum of the *sinistralateral* and *dextralateral* relationships is the reactant relationship. These relationships are directly specified in the database. Note that *tint* is not a proper *edge* coloring as two adjacent *edges* can have identical colors.

titre (titer)   The reacting strength of a standard solution, usually expressed as the weight (mass) of the substance equivalent to 1 cm^3 of the solution.

token   See *term*.

tolerance approach   An approach to sensitivity analysis in linear programming that expresses the common range that parameters can occupy while preserving the character of the solution. In particular, suppose B is an optimal basis and rim data changes by (Db, Dc). The tolerance for this is the maximum value of t for which B remains optimal as long as $|Db_i| \leq t$ for all i and $|Dc_j| \leq t$ for all j. The tolerance for the basis, B, can be computed by simple linear algebra, using tableau information.

tolerances   Small positive values to control elements of a computer implementation of an *algorithm*. When determining whether a value, v, is nonnegative, the actual test is $v > -t$, where t is an absolute tolerance. When comparing two values to determine if $u \geq v$, the actual test is

$$u - v \leq t_a + t_r |v|,$$

where t_a is the absolute tolerance (as above), and t_r is the relative tolerance (some make the relative deviation depend on u as well as on v, such as the sum of magnitudes, $|u| + |v|$). Almost every MPS has a tolerance for every action it takes during its progression. In particular, one zero tolerance is not enough. One way to test feasibility is usually one that is used to determine an acceptable pivot element. In fact, the use of tolerances is a crucial part of an MPS, including any presolve that would fix a variable when its upper and lower bounds are sufficiently close (i.e., within some tolerance). A tolerance is *dynamic* if it can change during the *algorithm*. An example is that a high tolerance might be used for line search early in an *algorithm*, reducing it as the sequence gets close to an optimal solution. The Nelder-Mead simplex method illustrates how tolerances might change up and down during the *algorithm*.

Another typical tolerance test applies to residuals to determine if x is a solution to $Ax = b$.

In this case, the residual is $r = Ax - b$, and the test has the form:

$$\|r\| \leq t_a + t_r \|b\|,$$

where $\| \ \|$ is some norm.

topological invariant   A quantity enjoyed by a *topological space* which is invariant with respect to *homeomorphisms*.

topological motif   A subnetwork of the biochemical network invariant under topological transformation.

Comment: See also *biochemical, chemical, dynamical, functional, kinetic, mechanistic, phylogenetic, regulatory,* and *thermodynamic motives*.

topological sort   This sorts the nodes in a network such that each arc, say kth, has $\text{Tail}(k) < \text{Head}(k)$ in the renumbered *node* indexes. This arises in a variety of combinatorial optimization problems, such as those with precedence constraints. If the *nodes* cannot be topologically sorted, the network does not represent a partially ordered set. This means, for example, there is an inconsistency in the constraints, such as jobs that cannot be sequenced to satisfy the asserted precedence relations.

topological space   A pair $(X, \tau(X))$ where X is a set and $\tau(X)$ its *topology*. Different choices of the *topology* $\tau(X)$ of a space X correspond to different topological structures on X. Examples: If (X, d) is a *metric space*, then we can define $U \in \tau(X)$ if and only if for all $x \in U$ there exists an open ball $B_x^r = \{y \in X : d(x, y) < r\}$ such that $x \in B_x^r \subset U \subset X$. This is called the *metric topology of* (X, d).

On any set X we can define the *trivial topology* $\tau(X) = \{\emptyset, X\}$ and the *discrete topology* in which $\tau(X)$ is the set of all subsets of X (so that any subset of X is open in the discrete topology).

topological transformation   A one-to-one correspondence between the points of two geometric figures A and B which is continuous in both directions. If one figure can be transformed into another by a topological transformation, the two figures are said to be topologically equivalent.

Comment: This is one of two standard definitions. Common examples are smooth deformation of a triangle into a circle; a sphere into a beaker (a cup without a handle); and a trefoil knot into a circle. In the last case, the transformation is allowed to cut the perimeter of the figure so long as the cut ends are rejoined in their original manner. Note that expansions and contractions of the network are not topological transformations.

topology (1) (of a network) The *topology* of a network $G(\mathcal{V}, \mathcal{E})$ is the set of *nodes* $\mathcal{V}$ and their incidence relations in the network, $\mathcal{E}$.

Comment: Specified here are two particular topological properties which are to remain invariant under a *topological transformation*, such as a continuous deformation. The standard definition requires only that the *edges* remain invariant to transformation. This is perfectly reasonable for networks derived from mathematics, but does not fit the biological case as well. Many networks will include singleton *nodes* which are important, but whose *edges* are not yet known. Hence the standard definition is here augmented to cover this case as well. There are a number of other important senses of the word which are not directly relevant here.

(2) (on a set X) A set $\tau(X)$ of subsets of X, called *open sets*, which satisfy the following axioms:

(i.) the empty set $\emptyset$ and the whole space X are elements in $\tau(X)$;

(ii.) the intersection of a finite number of elements in $\tau(X)$ is still in $\tau(X)$; and

(iii.) the union of a (possibly infinite) family of elements in $\tau(X)$ is still in $\tau(X)$.

torque, T Sum of *moments of forces* not acting along the same line.

torsion tensor Let $\Gamma^{\alpha}_{\beta\mu}$ be a (linear) connection on a *manifold* M. The torsion of the connection is the tensor $T^{\alpha}_{\beta\mu} = \Gamma^{\alpha}_{\beta\mu} - \Gamma^{\alpha}_{\mu\beta}$. Despite the fact that the connection is not a tensor, the torsion is a tensor since nonhomogeneous terms in the transformation rules of connections cancel out.

total ion current (in mass spectrometry) (1) (after mass analysis) The sum of the separate ion currents carried by the different ions contributing to the spectrum. This is sometimes called the reconstructed ion current.

(2) (before mass analysis) The sum of all the separate ion currents for ions of the same sign before mass analysis.

totally unimodular matrix See *unimodular matrix*.

toxicity (1) Capacity to cause injury to a living organism defined with reference to the quantity of substance administered or absorbed, the way in which the substance is administered (inhalation, ingestion, topical application, injection) and distributed in time (single or repeated doses), the type and severity of injury, the time needed to produce the injury, the nature of the organism(s) affected and other relevant conditions.

(2) Adverse effects of a substance on a living organism defined with reference to the quantity of substance administered or absorbed, the way in which the substance is administered (inhalation, ingestion, topical application, injection) and distributed in time (single or repeated doses), the type and severity of injury, the time needed to produce the injury, the nature of the organism(s) affected, and other relevant conditions.

(3) Measure of incompatibility of a substance with life. This quantity may be expressed as the reciprocal of the absolute value of median lethal dose ($1/LD_{50}$) or concentration ($1/LC_{50}$).

trace element Any element having an average concentration of less than about 100 parts per million atoms (ppma) or less than 100 μg per g.

trajectory (in reaction dynamics) A path taken by a reaction system over a *potential-energy surface*, or a diagram or mathematical description that represents that path. A *trajectory* can also be called a reaction path.

transfer Movement of a component within a system or across its boundary. It may be expressed using different kinds of quantities, e.g., rates of change dQ/dt or $\Delta Q/\Delta t$. Examples are mass rate, dm_B/dt or $\Delta m_B/\Delta t$; substance rate, dn_B/dt or $\Delta n_B/\Delta t$.

transformation (1) (in chemistry) The conversion of a *substrate* into a particular product, irrespective of reagents or *mechanisms* involved. For example, the transformation of aniline $(C_6H_5NH_2)$ into N-phenylacetamide $(C_6H_5NHCOCH_3)$ may be effected by use of acetyl chloride or acetic anhydride or ketene. A transformation is distinct from a reaction, the full description of which would state or imply all the reactants and all the products.

(2) (in mathematics) See *function*.

transient (chemical) species Relating to a short-lived reaction *intermediate*. It can be defined only in relation to a timescale fixed by the experimental conditions and the limitations of the technique employed in the detection of the intermediate. The term is a relative one.

Transient species are sometimes also said to be metastable. However, this latter term should be avoided, because it relates a thermodynamic term to a kinetic property, although most transients are also thermodynamically *unstable* with respect to reactants and products.

transient phase (induction period) The period that elapses prior to the establishment of a *steady state*. Initially the concentration of a reactive intermediate is zero, and it rises to the steady-state concentration during the transient phase.

transition function See *manifold*.

transition state In theories describing *elementary reactions* it is usually assumed that there is a *transition state* of more positive molar Gibbs energy between the reactants and the products through which an assembly of atoms (initially composing the *molecular entities* of the reactants) must pass on going from reactants to products in either direction. In the formalism of transition state theory the transition state of an elementary reaction is that set of states (each characterized by its own geometry and energy) in which an assembly of atoms, when randomly placed there, would have an equal probability of forming the reactants or of forming the products of that elementary reaction. The *transition state* is characterized by one and only one imaginary frequency. The assembly of atoms at the *transition state* has been called an activated complex. (It is not a *complex* according to the definition in this compendium.)

It may be noted that the calculations of reaction rates by the *transition state* method and based on calculated *potential-energy surfaces* refer to the potential energy maximum at the *saddle point*, as this is the only point for which the requisite separability of transition state coordinates may be assumed. The ratio of the number of assemblies of atoms that pass through to the products to the number of those that reach the *saddle point* from the reactants can be less than unity, and this fraction is the transmission coefficient κ. (There are reactions, such as the gas-phase *colligation* of simple *radicals*, that do not require activation and which therefore do not involve a transition state.)

transition state theory A theory of the rates of *elementary reactions* which assumes a special type of equilibrium, having an equilibrium constant $K^{\ddagger}$, existing between reactants and activated complexes. According to this theory the rate constant is given by

$$k = (k_B T/h) K^{\ddagger}$$

where k_A is the Boltzmann constant and h is the Planck constant. The rate constant can also be expressed as

$$k = (k_B T/h) \exp(\Delta^{\ddagger} S^0 / R) \exp(-\Delta^{\ddagger} H^0 / RT)$$

where $\Delta^{\ddagger} S^0$, the entropy of activation, is the standard molar change of entropy when the activated complex is formed from reactants and $\Delta^{\ddagger} H^0$, the enthalpy of activation, is the corresponding standard molar change of enthalpy. The quantities E_a (the *energy of activation*) and $\Delta^{\ddagger} H^0$ are not quite the same, the relationship between them depending on the type of reaction. Also

$$k = (k_B T/h) - \exp(-\Delta^{\ddagger} G^0 / RT)$$

where $\Delta^{\ddagger} G^0$, known as the *Gibbs energy of activation*, is the standard molar Gibbs energy change for the conversion of reactants into activated complex. A plot of standard molar

Gibbs energy against a reaction coordinate is known as a Gibbs-energy profile; such plots, unlike *potential-energy profiles*, are temperature-dependent.

In principle, the expressions for k above must be multiplied by a transmission coefficient, κ, which is the probability that an activated complex forms a particular set of products rather than reverting to reactants or forming alternative products.

It is to be emphasized that $\Delta^{\ddagger}S^0$, $\Delta^{\ddagger}H^0$, and $\Delta^{\ddagger}G^0$ occurring in the former three equations are not ordinary thermodynamic quantities, since one *degree of freedom* in the activated complex is ignored.

Transition-state theory has also been known as the absolute rate theory, and as activated-complex theory, but these terms are no longer recommended.

transition structure   A *saddle point* on a *potential-energy surface*. It has one negative force constant in the harmonic force constant matrix. See also *transition state*.

translation   The action of a group $(V, +)$ regarded as a group of transformations on V itself through the action $T : V \times V \to V$ given by

$$T : (T, v) \mapsto v + T.$$

transport equation   Let $U \subset \mathbb{R}^n$ be open and $u : U \times \mathbb{R} \to \mathbb{R}$. The (linear) *transport equation* for u is

$$u_t + \sum_{i=1}^{n} b^i u_{x_i} = 0.$$

transportation problem   Find a flow of least cost that ships from supply sources to consumer destinations. This is a bipartite network, $N = [S^*T, A]$, where S is the set of sources, T is the set of destinations, and A is the set of arcs. In the standard form, N is bi-complete (A contains all arcs from S to T), but in practice networks tend to be sparsely linked. Let $c(i, j)$ be the unit cost of flow from i in S to j in T, $s(i) =$ supply at ith source, and $d(j) =$ demand at jth destination. Then, the problem is the linear program

$$\text{Minimize} \sum_{ij} \{c(i, j)x(i, j) : i \text{ in } S, j \text{ in} T\}$$

where $x \geq 0$,

$$\sum_{j} \{x(i, j) : j \text{ in } T\} \leq s(i) \text{ for all } i \text{ in } S,$$

$$\sum_{i} \{x(i, j) : i \text{ in } S\} \geq d(j) \text{ for all } j \text{ in } T.$$

The decision variables (x) are called flows, and the two classes of constraints are called supply limits and *demand requirements*, respectively. (Some authors use equality constraints, rather than the inequalities shown.) An extension is the capacitated transportation problem, where the flows have bounds $x \leq U$.

transpose (of a matrix $[a_{ij}]$)   The matrix $[a_{ji}]$.

transposition theorem   Same as a *theorem of the alternative*.

transshipment problem   This is an extension of the *transportation problem* whereby the network is not bipartite. Additional nodes serve as transshipment points, rather than providing supply or final consumption. The network is $N = [V, A]$, where V is an arbitrary set of nodes, except that it contains a nonempty subset of supply nodes (where there is external supply) and a nonempty subset of demand nodes (where there is external demand). A is an arbitrary set of arcs, and there could also be capacity constraints.

traveling salesman problem (TSP)   Given n points and a cost matrix, $[c(i, j)]$, a tour is a permutation of the n points. The points can be cities, and the permutation the visitation of each city exactly once, then returning to the first city (called   home). The cost of a tour, $\langle i_1, i_2, ..., i_{n-1}, i_n, i_1 \rangle$, is the sum of its costs:

$$c(i_1, i_2) + c(i_2, i_3) + ... + c(i_{n-1}, i_n) + c(i_n, i_1),$$

where $(i_1, i_2, ..., i_n)$ is a permutation of $\{1, ..., n\}$. The TSP is to find a tour of minimum total cost. The two common integer programming formulations are:

ILP: $\min \sum_{ij} c_{ij} x_{ij} : x \in P, \ x_{ij} \in \{0, 1\}$

Subtour elimination constraints: $\sum_{i,j \in V} x_{ij} \leq |V| - 1$ for $\emptyset \neq V \subset \{1, ..., n\}$ ($V \neq \{1, ..., n\}$)

where

$$x_{ij} = \begin{cases} 1 & j \text{ follows city } i \text{ in tour} \\ 0 & \text{otherwise} \end{cases}$$

QAP: $\min \sum_{ij} c_{ij} \left(\sum_{k=1}^{n-1} x_{ik} x_{jk+1} + x_{in} x_{j1} \right)$:

$$x \in P, \ x_{ij} \in \{0, 1\},$$

where

$$x_{ij} = \begin{cases} 1 & \text{if tour has city } i \text{ in position } j \\ 0 & \text{otherwise} \end{cases}$$

In each formulation, P is the assignment polytope. The *subtour elimination constraints* in ILP eliminate assignments that create cycles.

For example, in subtours $(1 \rightarrow 2 \rightarrow 3 \rightarrow 1)$ (length 3) and $(4 \rightarrow 5 \rightarrow 4)$ (length 2), the first subtour is eliminated by $V = \{1, 2, 3\}$, which requires $x_{12} + x_{23} + x_{31} \le 2$. The second subtour is eliminated by $V = \{4, 5\}$, which requires $x_{45} + x_{54} \le 1$.

tree   A connected graph containing no cycles.

triangle inequality   A property of a distance function: $f(x, y) \le f(x, z) + f(z, y)$ for all x, y, z.

triangular matrix   A square *matrix A*, is called *upper triangular* if all elements are zero below the main diagonal, i.e., $A(i, j) = 0$ for $i > j$. It is called *lower triangular* if its *transpose* is upper triangular. We sometimes call a matrix *triangular* if it is either lower or upper triangular.

triangulation/mesh   Given a bounded domain $\Omega \subset \mathbb{R}^n$ with piecewise smooth boundary a triangulation/mesh Ω_h of Ω is a finite set $\{K_i\}_{i=1}^M$, $M \in \mathbb{N}$, of piecewise smooth open subsets of Ω, called *cells*, such that

(i.) the interior of the closure of each cell is the cell itself.

(ii.) the union $\bigcup_{K \in \Omega_h} \bar{K}$ coincides with $\bar{\Omega}$ and $K_i \cap K_j = \emptyset$, if $i \ne j$ (open partition property).

(iii.) The intersection of the closures of any two cells is either empty or a vertex, edge, face, etc., of both.

This means that a triangulation constitutes a nondegenerate *cellular decomposition* of Ω. Special types of meshes are *simplicial meshes*, for which all the cells are n-simplices. In two dimensions the cells of *quadrilateral meshes* have four, possibly curved, edges each. Their three-dimensional counterparts are *hexaedral meshes*, whose cells are bricks (with curved faces and edges). In a straightforward fashion the concept can be generalized to the notion of a triangulation of a compact piecewise smooth manifold (with or without boundary). In the case of *adaptive refinement* it is often desirable to relax the above requirements by admitting *hanging nodes*. These are vertices of some cells that lie in the interior of edges of other cells.

triple point   The point in a one-component system at which the temperature and pressure of three phases are in equilibrium. If there are p possible phases, there are $p!/(p-3)!3!$ triple points. Example: In the sulfur system four possible triple points (one metastable) exist for the four phases comprising rhombic S (solid), monoclinic S (solid), S (liquid), and S (vapor).

triplet state   A state having a total electron spin quantum number of 1.

triprismo-   An affix used in names to denote six atoms bound into a triangular prism.

trivial bundle   A bundle $(B, M, \pi; F)$ which has a global trivialization so that the total space is *diffeomorphic* to the Cartesian product $B \simeq M \times F$. *Trivial bundles* always allow global sections, and they are the local model of all bundles.

tropopause   The region of the atmosphere which joins the *troposphere* and *stratosphere*, and where the decreasing temperature with altitude, characteristic of the troposphere ceases, and the temperature increase with height which is characteristic of the stratosphere begins.

troposphere   The lowest layer of the atmosphere, ranging from the ground to the base of the *stratosphere (tropopause)* at 10–15 km of altitude depending on the latitude and meteorological conditions. About 70% of the mass of the atmosphere is in the troposphere. This is where most of the weather features occur and where the chemistry of the reactive anthropogenic species released into the atmosphere takes place.

Trotter product formula   If A and B are *self-adjoint* operators and $A + B$ is *essentially self-adjoint*, then

$$\lim_{n \to \infty} (e^{itA/n} e^{itB/n})^n = e^{i(A+B)t}.$$

true value (in analysis), τ   The value that characterizes a quantity perfectly in the conditions that exist when that quantity is considered. It is an ideal value which could be arrived at only if all causes of measurement error were eliminated, and the entire population was sampled.

truncated gradient   Projection of the gradient of a function, f (in C^1) on a box, $[a, b]$, to put zeros in coordinates if the sign of the partial derivative is negative at the lower bound or it is positive at the upper bound. This yields a feasible direction.

trust region method   The iteration is defined as $x' = x + p$, where p is the (complete) change (no separate step size), determined by a subproblem of the form

$$\text{Max } F(p) : \|p\| \leq D,$$

where F depends on the iterate and is an approximation of the change in objective function value. The particular norm and the magnitude of D determine the set of admissible change values (p), and this is called the *trust region*. A common choice of F is the quadratic form using the Taylor expansion about the current iterate, x, as

$$F(p) = \text{grad}_f(x)p + p'[H_f(x)]p/2.$$

Using the Euclidean norm and applying the Lagrange multiplier rule to the subproblem yields p from the equation

$$[H_f(x) - uI]p = -\text{grad}_f(x) \text{ for some } u \geq 0.$$

Note that for $u = 0$, the iteration is Newton's method, and for very large u, the iteration is nearly Cauchy's steepest ascent.

tub conformation   A *conformation* (of symmetry group D_{2d}) of an eight-membered ring in which the four atoms forming one pair of diametrically opposite bonds in the ring lie in one plane and all other ring atoms lie to one side of that plane. It is analogous to the boat conformation of cyclohexane.

Tung distribution (of a macromolecular assembly)   A continuous distribution with the differential mass-distribution function of the form:

$$f_w(x)dx = abx^{b-1} \exp(-ax^b)dx$$

where x is a parameter characterizing the chain length, such as *relative molecular mass* or *degree of polymerization* and a and b are positive adjustable parameters.

tunneling   The process by which a particle or a set of particles crosses a barrier on its *potential-energy surface* without having the energy required to surmount this barrier. Since the rate of tunneling decreases with increasing reduced mass, it is significant in the context of *isotope effects* of hydrogen isotopes.

tuple   A collection of elements which satisfy some relation r. A tuple is delimited by parentheses $((x, y))$, and is usually written with the relation as the *functor* of the tuple; thus $r(x, y)$.

Comment: Note that a sequence is *not* equivalent to a tuple. The functor is omitted when it is absolutely clear what relation it satisfies. See also *bag, list, relation, sequence,* and *set.*

Turing pattern   A.M. Turing was the first person who proposed a mechanism for pattern formation in a spatially homogeneous system to involve chemical reactions and diffusion of two species. The pattern formation is due to *diffusion-driven instability.*

U

unbounded mathematical program The objective is not bounded on the feasible region (from above, if maximizing; from below, if minimizing). Equivalently, there exists a sequence of feasible points, say $\{x^k\}$ for which $\{f(x^k)\}$ diverges to infinity (minus infinity, if minimizing).

unconstrained mathematical program One with no constraints (can still have X be a proper subset of $\mathbb{R}^n$, such as requiring x to be integer-valued.)

unconstrained optimization Taken literally, this is an unconstrained mathematical program. However, this phrase is also used in a context that X could contain the strict interior, with constraints of the form $g(x) < 0$, but the mathematical program behaves as unconstrained. This arises in the context of some *algorithm* design, as the solution is known to lie in the interior of X, such as with the barrier function.

uncountably infinite set An infinite set which is not denumerably so. See also *cardinality, countable set, denumerably infinite set, finite set*, and *infinite set*.

undirected edge See *edge*.

unified atomic mass unit Non-SI unit of mass (equal to the atomic mass constant), defined as one twelfth of the mass of a carbon-12 atom in its ground state and used to express masses of atomic particles, $u = 1.660\ 5402(10) \times 10^{-27}$ kg.

uniformly bounded Referring to a family F of functions such that the same bound holds for all functions in F. For example,

$$f(x) \leq \mu$$

with the same μ, for all $f \in F$.

unimodal function A function which has one mode (usually a maximum, but could mean a minimum, depending on context). If f is defined on the interval $[a, b]$, let x^* be its mode. Then, f strictly increases from a to x^* and strictly decreases from x^* to b (reverse the monotonicity on each side of x^* if the mode is a minimum). (For line search methods, like Fibonacci, the mode could occur in an interval, $[a^*, b^*]$, where f strictly increases from a to a^*, is constant (at its global max value) on $[a^*, b^*]$, then strictly decreases on $[b^*, b]$.)

unimodular matrix A nonsingular matrix whose determinant has magnitude 1. A square matrix is totally unimodular if every nonsingular submatrix from it is unimodular. This arises in (linear) integer programming because it implies a basic solution to the LP relaxation is integer-valued (given integer-valued right-hand sides), thus obtaining a solution simply by a simplex method. An example of a totally unimodular matrix is the node-arc incidence matrix of a network, so basic solutions of network flows are integer-valued (given integer-valued supplies and demands).

unimolecular See *molecularity*.

unisolvence A set of functionals on a finite dimensional *vector space V_h* is called *unisolvent*, if it provides a basis of the dual space of V_h. Unisolvence is an essential property of *degrees of freedom* in the finite element method.

unit An *identity* element.

unit circle A circle of radius 1. Usually, the term refers to the circle of radius 1 and center 0 in the complex plane ($\{z : |z| = 1\}$).

unitary group An automorphism α : $\mathbb{C}^m \to \mathbb{C}^m$ is called *unitary* if, once a basis E_i has been chosen in $\mathbb{C}^m$, the matrix U_i^j representing the automorphism by $\alpha(E_j) = U_i^j E_j$ satisfies $U^{-1} = U^\dagger$, i.e., if the inverse of U coincides with the *transpose* of the complex conjugated matrix.

The group of all such matrices is called the *unitary group*, and it is denoted by $U(m)$. The subgroup of unitary matrices with $\det U = 1$ is denoted by SU(m).

unitary matrix   A nonsingular matrix whose Hermitian adjoint equals its inverse (same as *orthogonal* for real-valued matrices). See *self-adjoint operator*.

univariate optimization   A mathematical program with a single variable.

universal set   The set containing all sets, or sets of interest; denoted $\mathcal{U}$. See also *universe of discourse*.

Comment: In databases, one frequently speaks of a *universe of discourse*, the set of all terms, facts, relations, and functions used in reifying the database's model of its world (its *domain model*). For many purposes the two are equivalent.

universal Turing machine   A *universal Turing machine* (UTM) is a discrete automaton that executes a computation $\mathcal{C}$. It has a read-write head that reads symbols from and writes them to an unbounded but finite, immutable memory. The memory stores symbols from a finite symbol set

$$\Sigma = \Sigma_{\mathcal{I}} \cup \Sigma_{\mathcal{O}} \cup \Sigma_{\mathcal{C}},$$

where $\overline{\Sigma}_{\mathcal{I}} = 2$, $\overline{\Sigma}_{\mathcal{O}} = 1$, and $\Sigma_{\mathcal{C}}$ are the symbols internal to the computation. At each step the automaton assumes one of a finite number of discrete states, $s_i, s_i \in \mathcal{S}$. A computation $\mathcal{C}$ is defined as a set of tuples each of the form $(s_i, \sigma_{i,\mathcal{I}'}, s_{i+1}, \sigma_{i,\mathcal{O}'}, a_i)$, where s_i is the automaton's state at the ith step, $\sigma_{i,\mathcal{I}'}$ the symbol read into the automaton at that step ($\overline{\Sigma}_{i,\mathcal{I}'} = \overline{\Sigma}_{i,\mathcal{O}'} = 1$), s_{i+1} is the new state the automaton assumes upon completion of step i (which will be its state as it commences step $i + 1$), $\sigma_{i,\mathcal{O}'}$ is the symbol output at step i, and a_i is the action altering the position of the tape in the head that is performed at the end of that step (move it left, right, or nowhere). The values for each σ and $\int$ persist in the automaton long enough for it to execute each step. One of the symbols input to $\mathcal{C}$ refers to an emulation of a particular Turing machine stored in the UTM memory. This emulation is the set of all tuples $(s_i, \sigma_{i,\mathcal{I}'}, s_{i+1}, \sigma_{i,\mathcal{O}'}, a_i)$ for that machine and forms the *algorithm*. The *algorithm* may be deterministic or nondeterministic, but the machine executes it nonstochastically.

Comment: This definition differs slightly from Minsky's (*cf.* M.L. Minsky, *Computation: Finite and Infinite Machines*, Prentice-Hall, Englewood Cliffs, NJ, 1967). See *Post production system* and *von Neumann machine*.

universe of discourse   The nonempty set, $\mathcal{U}_d$, of all possible constant terms of a program. More generally, the area of nature, thought, or existence described by a program or set of programs.

Comment: For many purposes, this is equivalent to the *universal set*. See *universal set*.

unreactive   Failing to react with a specified *chemical species* under specified conditions. The term should not be used in place of *stable*, since a relatively more stable species may nevertheless be more *reactive* than some reference species toward a given reaction partner.

unstable   The opposite of *stable*, i.e., the *chemical species* concerned has a higher molar Gibbs energy than some assumed standard. The term should not be used in place of *reactive* or *transient*, although more reactive or transient species are frequently also more unstable.

Very unstable chemical species tend to undergo exothermic *unimolecular* decompositions. Variations in the structure of the related chemical species of this kind generally affect the energy of the *transition states* for these decompositions less than they affect the stability of the decomposing chemical species. Low stability may therefore parallel a relatively high rate of unimolecular decomposition.

upper semicontinuity (or upper semicontinuous [USC])   Suppose $\{x^k\} \rightarrow x$.
Of a function, $\limsup f(x^k) = f(x)$.

Of a point-to-set map, let $N_e[S]$ be a *neighborhood* of the set S. For each $e > 0$, there exists K such that for all $k > K$, $A(x^k)$ is a subset of $N_e[A(x)]$. Here is an example of what can go wrong. Consider the feasibility map with

$$g(x) = \begin{cases} (x + \sqrt{2})^2 - 1 & \text{if } x < 0 \\ \exp\{-x\} & \text{if } x \geq 0 \end{cases}$$

Note g is continuous and its level set is $[-\sqrt{2} - 1, -\sqrt{2} + 1]$. However, for any $b > 0$, $\{x : g(x) \leq b\} = [-\sqrt{2} - \sqrt{1 + b}, -\sqrt{2} + \sqrt{1 + b}]/[-\log b, \infty)$, which is not bounded. The map fails to be (USC) at 0 due to the lack of stability of its feasibility region when perturbing its right-hand side (from above).

upper triangular matrix   A square matrix, A, such that $A(i, j) = 0$ for $i \geq j$.

utility function   A measure of benefit, used as a maximand in economic models.

V

valid inequality An inequality constraint added to a relaxation that is redundant in the original mathematical program. An example is a linear form, $ax \leq b$, used as a cutting plane in the LP relaxation of an integer program.

A linear form that is a facet of the integer polyhedron.

value iteration This is an *algorithm* for infinite horizon [stochastic] dynamic programs that proceeds by successive approximation to satisfy the fundamental equation

$$F(s) = \text{Opt} \{r(x, s) + a \sum_{s'} P(x, s, s')F(s')\},$$

where a is a discount rate. The successive approximation becomes the DP forward equation. If $0 < a < 1$, this is a fixed point, and Banach's theorem yields convergence because then "Opt" is a contraction map. Even when there is no discounting, policy iteration can apply.

value (of a quantity) Magnitude of a particular quantity generally expressed as a unit of measurement multiplied by a number.

variable metric method Originally referred to as the Davidon-Fletcher-Powell (DFP) method, this is a family of methods that choose the direction vector in unconstrained optimization by the subproblem: d^* in argmax $\{\text{grad} f(x)d : \|d\| = 1\}$, where $\|d\|$ is the vector norm (or metric) defined by the quadratic form, $d'Hd$. With H symmetric and positive definite, the constraint $d'Hd = 1$ restricts d by being on a circle, i.e., equidistant from a stationary point, called the center (the origin in this case). By varying H, as in the DFP update, to capture the curvature of the objective function, f, we have a family of ascent algorithms. Besides DFP, if one chooses $H = I$, we

have Cauchy's steepest ascent. If f is concave and one chooses H equal to the negative of the inverse Hessian, we have the modified Newton's method.

variable upper bound (VUB) A constraint of the form: $x_i \leq x_j$.

variational calculus An approach to solving a class of optimization problems that seek a *functional* (y) to make some integral function (J) an extreme. Given F in C^1, the classical unconstrained problem is to find y in C^1 to minimize (or maximize) the following function:

$$J(y) = \int_{x_0}^{x_1} F(x, y, y')dx.$$

An example is to minimize arc length, where $F = \sqrt{(1 + y'^2)}$. Using the Euler-Lagrange equation, the solution is $y(x) = ax + b$, where a and b are determined by *boundary conditions:* $y(x_0) = y_0$ and $y(x_1) = y_1$.

If constraints take the form $G(x, y, y') = 0$, this is called the problem of Lagrange; other forms are possible.

variational crime Consider a linear variational problem $a(u, v) = f(v), v \in V, f \in V', a : V \times V \to \mathbb{C}$ a *sesqui-linear* form, V a Banach space. In two ways a discretization based on a *finite element space* V_h can go beyond the usual *Galerkin framework*:

(i.) The *sesquilinear* form a and the linear form f may be replaced with mesh-dependent approximations $a_h : V_h \times V_h \to \mathbb{C}$ and $f_h : V_h \to \mathbb{C}$. This happens, for instance, when numerical quadrature is employed for the evaluation of integrals.

(ii.) The finite element space V_h might be *non-conforming* with respect to V, that is $V_h \not\subset V$.

Whether these *variational crimes* still lead to a meaningful discrete problem can be checked by means of *Strang's lemmas*.

variational inequality Let $F : X \to \mathbb{R}^n$. The *variational inequality* problem is to find x in X such that $F(x)(y - x) \geq 0$ for all y in X. This includes the complementarity problem.

vector   (1) A directed line segment in $\mathbb{R}^2$ or $\mathbb{R}^3$, determined by its magnitude and direction.

(2) An element of a *vector space*.

vector bundle   A bundle $(B, M, \pi; V)$ with a *vector space* V as a *standard fiber* and such that the *transition functions* act on V by means of *linear transformations*. *Vector bundles* always allow global sections.

vector field   A vector-valued function V, defined in a region D (usually in $\mathbb{R}^3$). The vector $V(p)$, assigned to a point $p \in D$, is required to have its initial point at p.

vector product   For two vectors $x = (x_1, x_2, x_3)$ and $y = (y_1, y_2, y_3)$ in $\mathbb{R}^3$, the vector

$$x \times y = (x_2 y_3 - x_3 y_2, x_3 y_1 - x_1 y_3, x_1 y_2 - x_2 y_1).$$

vector space   A set closed under addition and scalar multiplication (by elements from a given field). One example is $\mathbb{R}^n$, where addition is the usual coordinate-wise addition, and scalar multiplication is $t(x_1, ..., x_n) = (tx_1, ..., tx_n)$. Another *vector space* is the set of all $m \times n$ matrices. If A and B are two matrices (of the same size), so is $A + B$. Also, tA is a matrix for any scalar, t in $\mathbb{R}$. Another *vector space* is the set of all functions with domain X and range in $\mathbb{R}^n$. If f and g are two such functions, so are $f + g$ and tf for all t in $\mathbb{R}$. Note that a *vector space* must have a zero since we can set $t = 0$. See also *module*.

vehicle routing problem (VRP)   Find optimal delivery routes from one or more depots to a set of geographically scattered points (e.g., population centers). A simple case is finding a route for snow removal, garbage collection, or street sweeping (without complications, this is akin to a shortest path problem). In its most complex form, the VRP is a generalization of the TSP, as it can include additional time and capacity constraints, precedence constraints, plus more.

velocity, v, c   Vector quantity equal to the derivative of the position vector with respect to time (symbols, u, v, w for components of c).

vertex   See *node*.

vertex cover   Given a graph, $G = [V, E]$, a *vertex cover* is a subset of V, say C, such that for each edge (u, v) in E, at least one of u and v is in C. Given weights, $\{w(v)\}$ for v in V, the *weight* of a *vertex cover* is the sum of weights of the *nodes* in C. The *minimum weight vertex cover problem* is to find a *vertex cover* whose weight is minimum.

vertical automorphism   A *bundle morphism* $\Phi : B \to B$ on a bundle $(B, M, \pi; F)$ which projects over the identity, i.e., such that $\pi \equiv \mathrm{id}_M \circ \pi = \pi \circ \Phi$. Locally a vertical morphism is of the following form

$$\begin{cases} x' = x \\ y' = Y(x, y). \end{cases}$$

vertical vector field   A *vector field* Ξ over a bundle $(B, M, \pi; F)$ which projects over the zero vector of M. Locally, it is expressed as

$$\Xi = \xi^i(x, y)\partial_i$$

where $(x^\mu; y^i)$ are fibered coordinates. The *flow* of a *vertical vector field* is formed by *vertical automorphisms*.

viscosity   See *dynamic viscosity*.

voltage (in electroanalysis)   The use of this term is discouraged, and the term *applied potential* should be used instead, for nonperiodic signals. However, it is retained here for sinusoidal and other periodic signals because no suitable substitute for it has been proposed.

von Neumann machine (vNM)   A device that stores a program specifying an *algorithm* or *heuristic* in memory; includes separable arithmetic and logic processing units and input/output facilities and executes a discrete computation on the input data using the stored program. It assumes a finite number of distinct internal states and operates on a finite number of symbols Σ. It may or may not store the input and output, as desired. For convenience allow execution to be either sequential on a single processor or sequential within each of a set of processors joined together by various schemes for message passing and storage. The machine is a nonstochastic device executing deterministic or nondeterministic algorithms. See *universal Turing machine* and *Post production system*.

wall-coated open-tubular (WCOT) column (in chromatography) A column in which the liquid stationary phase is coated on the essentially unmodified smooth inner wall of the tube.

warehouse problem   The manager of a warehouse buys and sells the stock of a certain commodity, seeking to maximize profit over a period of time, called the *horizon*. The warehouse has a fixed capacity, and there is a holding cost that increases with increasing levels of inventory held in the warehouse (this could vary period to period). The sales price and purchase cost of the commodity fluctuate. The warehouse is initially empty and is required to be empty at the end of the horizon. This is a variation of the production scheduling problem, except demand is not fixed. (Level of sales is a decision variable, which depends on whether cost is less than price.)

Let

$$
\begin{aligned}
x(t) &= \text{level of production in period} t \\
&\quad \text{(before sales)} \\
y(t) &= \text{level of inventory at the end of} \\
&\quad \text{period } t \\
z(t) &= \text{level of sales in period } t \\
W &= \text{warehouse capacity} \\
h(t, u) &= \text{holding cost of inventory } u \\
&\quad \text{from period } t \text{ to } t + 1 \\
p(t) &= \text{production cost (per unit of production)} \\
s(t) &= \text{sales price (per unit)} \\
T &= \text{horizon}
\end{aligned}
$$

Then, the mathematical program is:

$$
\text{Minimize} \sum_t h(t, y(t)) + px - sz :
$$

$$
x, y, z \geq 0; \; y(0) = y(T) = 0;
$$

$$
y(t) = y(t - 1) + x(t) - z(t) \Leftarrow W
$$

for $t = 1, ..., T$.

wash out (in atmospheric chemistry)   The removal from the *atmosphere* of gases and sometimes particles by their solution in or attachment to raindrops as they fall.

wave equation   Let $U \subset \mathbb{R}^n$ open and $u : U \times \mathbb{R} \to \mathbb{R}$. The (linear) *wave equation* for u is

$$
u_{tt} - \Delta u = 0.
$$

The (nonlinear) *wave equation* for u is

$$
u_{tt} - \Delta u = f(u).
$$

wave function (state function), Ψ, ψ, ϕ   The solution of the *Schrödinger equation*, eigenfunction of the Hamiltonian operator.

wave height (electrochemical)   The *limiting current* of an individual wave, frequently expressed in arbitrary units for convenience.

wave number, $\sigma \tilde{\nu}$ The reciprocal of the *wavelength* λ, or the number of waves per unit length along the direction of propagation. Symbols $\tilde{\nu}$ in a vacuum, σ in a medium.

wavelength, λ   Distance in the direction of propagation of a periodic wave between two successive points where at a given time the phase is the same.

wavelength converter   Converts *radiation* at one wavelength to radiation at another detectable wavelength or at a wavelength of improved *responsivity* of the detector. The classical wavelength converter consists of a screen of luminescent material that absorbs radiation and radiates at a longer wavelength. Such materials are often used to convert ultraviolet to visible radiation for detection by conventional phototubes. In X-ray spectroscopy, a converter that emits optical radiation is called a *scintillator*. In most cases wavelength conversion is from short to long wavelength, but in the case of conversion of long to short wavelength the process is sometimes called upconversion. Wavelengths of coherent sources can be converted using nonlinear optical techniques. A typical example is frequency doubling.

wavelength dispersion (in x-ray emission spectroscopy) Spatial separation of characteristic x-rays according to their wavelengths.

wavelength-dispersive x-ray fluorescence analysis A kind of *x-ray fluorescence analysis* involving the measurement of the wavelength spectrum of the emitted *radiation* e.g., by using a diffraction grating or crystal.

wavelength error (in spectrochemical analysis) The error in *absorbance* which may occur if there is a difference between the (mean) wavelength of the radiation entering the sample cell and the indicated wavelength on the spectrometer scale.

wavelet An orthonormal basis of $L^2(\mathbb{R}^n)$. They are used as phase space localization methods. A typical example of a family of wavelets $\psi_{j,k}(x)$ in one dimension is given by

$$\psi_{j,k}(x) = 2^{-j/2}\psi(2^{-j}x - k)$$

$$= 2^{\frac{-j}{2}}\psi(\frac{x - 2^j k}{2^j}), \quad j, k \in \mathbb{Z},$$

where ψ is a smooth function with reasonable decay (say, $|\psi(x)| < C(1 + |x|)^{-(1+\epsilon)}$), and such that $\int \psi(x)dx = 0$. Wavelets bases provide unconditional bases for many classical function spaces, such as L^p spaces $1 < p < \infty$, Sobolev spaces W^s, Besov spaces $B_q^{p,s}$, Hölder spaces C^s, Hardy space H^1, and BMO space.

wavelet transformations Approximations that can have lower complexity than *fast Fourier transforms* (FFT), namely, $O(\log N)$ instead of $O(N \log N)$.

weak collision A collision between two molecules in which the amount of energy transferred from one to the other is not large compared to $k_B T$ (k_B is the *Boltzmann constant* and T the absolute temperature). See *strong collision*.

weak formulation To cast a boundary value problem for a partial differential operator into *weak form* means that it is stated as a variational problem over some Banach space V. The usual way to get this weak formulation from the *strong*

form of a boundary value problem is through integration by parts, using Green's formulas.

wedge projection A stereochemical projection, roughly in the mean plane of the molecule, in which bonds are represented by open wedges, tapering off from the nearer atom to the farther atom.

Weierstrass theorem See *Bolzano-Weierstrass theorem*.

weight, G Force of gravity acting on a body, $G = mg$, where m is its mass and g the acceleration of free fall.

weighted mean If in a series of observations a statistical weight (w_i) is assigned to each value, a weighted mean $\overline{x}_w$ can be calculated by the formula

$$\overline{x}_w = \frac{\sum w_i x_i}{\sum w_i}$$

Comment: Unless the weights can be assigned objectively, the use of the weighted mean is not normally recommended.

well-ordered set A partially ordered set $(A, \triangleright)$ such that for all $B \subset A$ (B non-empty) B has a *minimum*.

well posed Most people mean that the mathematical program has an optimal solution. Some mean just that it is feasible and not unbounded.

wet bulb temperature In *psychrometry*, the temperature of the sensor or the bulb of a thermometer in which a constantly renewed film of water is evaporating. The temperature of the water used to renew the film must be at the temperature of the gas.

wetting tension (or work of immersional wetting per unit area) The work done on a system when the process of *immersional wetting* involving unit area of phase β is carried out reversibly:

$$wW^{\alpha\beta\delta} = \gamma^{\beta\delta} - \gamma^{\alpha\beta}$$

where $\gamma^{\alpha\beta}$ and $\gamma^{\beta\delta}$ are the *surface tensions* between two bulk phases α, β and β, δ, respectively.

White's formula A closed DNA molecule modeled as a ribbon has a twist (Tw) with respect to its central line (the line of centroids) and a writhe (Wr) which represents the contortion of the central line in space. It was shown by J.H. White, that Tw + Wr = linking number, a topological invariant. White's formula relates DNA structure to knot theory (see also *DNA supercoil*; *cf.* J.H. White, *Am. J. Math.*, **91**, 693 (1969)).

wind rose A diagram designed to show the distribution of wind direction experienced at a given location over a considerable period of time. Usually shown in polar coordinates (distance from the origin being proportional to the probability of the wind direction being at the given angle usually measured from the north). Similar diagrams are sometimes used to summarize the average concentrations of a given pollutant seen over a considerable period of time as a function of direction from a given site (sometimes called a pollution rose).

wood horn A mechanical device that acts by absorption as a perfect photon trap.

wood lamp A term used to describe a low-pressure mercury arc. See *lamp*.

work, w, W Scalar produce of force, F, and position change, $d\,r$, $w = \int F \times d\,r$.

work hardening Opposite of *work softening*, in which shear results in a permanent increase of *viscosity* or consistency with time.

work of adhesion The work of adhesion per unit area, $w_A^{\alpha\beta\delta}$, is the work done on the system when two condensed phases α and β, forming an interface of unit area, are separated reversibly to form unit areas of each of the $\alpha\delta$- and $\beta\delta$-interfaces.

$$w_A^{\alpha\beta\delta} = \gamma^{\alpha\delta} + \gamma^{\beta\delta} - \gamma^{\alpha\beta}$$

where $\gamma^{\alpha\beta}$, $\gamma^{\alpha\delta}$, and $\gamma^{\beta\delta}$ are the *surface tensions* between two bulk phases α, β; α, δ and β, δ, respectively.

The work of adhesion as defined above, and traditionally used, may be called the work of separation.

work of cohesion per unit area Of a single pure liquid or solid phase α, w_C^{α} is the work done on the system when a column α of unit area is split, reversibly, normal to the axis of the column to form two new surfaces each of unit area in contact with the equilibrium gas phase.

$$w_C^{\alpha} = 2\gamma^{\alpha}$$

where γ^{α} is the *surface tension* between phase α and its equilibrium vapor or a dilute gas phase.

work softening The application of a finite shear to a system after a long rest may result in a decrease of *viscosity* or consistency. If the decrease persists when the shear is discontinued, this behavior is called work softening (or shear breakdown), whereas if the original viscosity or consistency is recovered this behavior is called thixotropy.

working electrode An electrode that serves as a transducer responding to the excitation signal and the concentration of the substance of interest in the solution being investigated, and that permits the flow of current sufficiently large to effect appreciable changes of bulk composition within the ordinary duration of a measurement.

working set Constraints believed to be active at a solution.

wormlike chain (in polymers) A hypothetical *linear macromolecule* consisting of an infinitely thin chain of continuous curvature; the direction of a curvature at any point is random. The model describes the whole spectrum of chains with different degrees of chain stiffness from rigid rods to random coils, and is particularly useful for representing stiff chains. In the literature this chain is sometimes referred to as a Porod-Kratky chain. Synonymous with continuously curved chain.

X

ξ- (xi-)   A symbol used to denote unknown configuration at a chiral center.

xanthophylls   A subclass of *carotenoids* consisting of the oxygenated carotenes.

xenobiotics   Manmade compounds with chemical structures foreign to a given organism.

xenon lamp   An intense source of ultraviolet, visible, and near-infrared light produced by electrical discharge in xenon under high pressure.

xerogel   A term used for the dried out open structures which have passed a *gel* stage during preparation (e.g., silica gel); and also for fried out compact *macromolecular* gels such as gelatin or rubber.

XPRESS-MP   A mathematical programming modeling system and solver.

XPS   See *photoelectron spectroscopy.*

x-radiation   Radiation resulting from the interaction of high-energy particles or photons with matter.

x-ray escape peak   In a gamma or x-ray spectrum, the peak due to the *photoelectric effect* in the detector and escape, from the sensitive part of the detector, of the x-ray photon emitted as a result of the photoelectric effect.

x-ray fluorescence   The emission of characteristic x-radiation by an atom as a result of the interaction of electromagnetic *radiation* with its orbital electrons.

x-ray fluorescence analysis   A kind of analysis based on the measurement of the energies and intensities of characteristic x-radiation emitted by a test portion during *irradiation* with electromagnetic radiation.

x-ray intensity   Essentially all x-ray measurements are made by photon counting techniques but the results are seldom converted to radiant flux or *irradiance* or *radiant exposure.* The term *photon flux* would be appropriate if the measurements were corrected for detector efficiency but this is seldom done for x-ray chemical analysis. Therefore, the term x-ray intensity, I, is commonly used and expressed as photons/unit time detected. Likewise the term relative x-ray intensity, I_r, is used to mean the intensity for the analyte in an unknown specimen divided by the intensity for a known concentration of the analyte element.

x-ray level   An electronic state occurring as the initial or final state of a process involving the absorption or emission of x-ray radiation. It represents a many electron state which, in the purely atomic case, has total *angular momentum* ($J = L + S$) as a well-defined quantum number.

x-ray photoelectron spectroscopy (XPS)   Any technique in which the sample is bombarded with x-rays and photoelectrons produced by the sample are detected as a function of energy. ESCA (Electron Spectroscopy for Chemical Analysis) refers to the use of this technique to identify elements, their concentrations, and their chemical state within the sample.

x-ray satellite   A weak line in the same energy region as a normal x-ray line. Another name used for weak features is non-diagram line. Recommendations as to the use of these two terms have conflicted. Using the term *diagram line* as defined here, non-diagram line may well be used for all lines with a different origin. The majority of these lines originate from the dipole-allowed de-excitation of multiply ionized or excited states, and are called multiple-ionization satellites. A line where the initial state has two vacancies in the same shell, notably the K-shell, is called a hypersatellite. Other mechanisms leading to weak spectral features in x-ray emission are, e.g., resonance emission, the radiative auger effect, magnetic dipole, and electric quadrupole transitions and, in metals, plasmon excitation. Atoms with open electron shells, i.e., transition metals, lanthanides, and actinides, show a splitting of certain x-ray lines due to the electron interaction involving this open shell. Structures originating in all these ways as well as structures in the valence band of molecules and solid chemical compounds have in the past been given satellite designations.

x-ray spectroscopy   Consists of three steps:

(i.)   excitation to produce emission lines characteristic of the elements in the material,

(ii.)   measurement of their intensity, and

(iii.)   conversion of *x-ray intensity* to concentration by a calibration procedure which may include correction for *matrix effects*.

a constant. The value given corresponds to the Gregorian calendar year ($a = 365.2425$ d).

yield, Y (in biotechnology)   Ratio expressing the efficiency of a mass conversion process. The yield coefficient is defined as the amount of *cell* mass (kg) or product formed (kg, mol) related to the consumed *substrate* (carbon or nitrogen source or oxygen in kg or moles) or to the intracellular ATP production (moles).

yield stress   The *shear stress* σ_0 or τ_0 at which yielding starts abruptly. Its value depends on the criterion used to determine when yielding occurs.

Young's inequality   Let $p, q, r \geq 1$ and $1/p + 1/q + 1/r = 2$. Let $f \in L^p(\mathbb{R}^n)$, $g \in L^q(\mathbb{R}^n)$, $h \in L^r(\mathbb{R}^n)$. Then

$$\left| \int_{\mathbb{R}^n} f(x)(g * h)(x)dx \right|$$
$$\leq C_{p,q,r,n} \|f\|_p \|g\|_q \|h\|_r.$$

Yukawa-Tsuno equation   A multiparameter extension of the *Hammett equation* to quantify the role of enhanced *resonance effects* on the reactivity of *meta-* and *para*-substituted benezene derivatives, e.g.,

$$\lg k = \lg k_0 + \rho[\sigma + r(\sigma^+ - \sigma)].$$

The parameter r gives the enhanced resonance effect on the scale $(\sigma^+ - \sigma)$ or $(\sigma - \sigma)$, respectively.

Yang-Mills theory   A *nonabelian gauge theory*. For a fixed compact *Lie group G* with *Lie algebra* **g**, the field in this theory is vector potential A, i.e., a *connection* 1-form on some principal G bundle. Let F_A be the curvature 2-form of the connection A. The fundamental Lagrangian in pure gauge theory is

$$L(A) = \frac{1}{2}|F_A||d^n x|,$$

where $|F_A|$ denotes the norm in the *Lie algebra*. From the variational principle we get the equations of motion, the *Yang-Mills equations*

$$d_A * F_A = 0,$$

where d_A is the *covariant derivative* with respect to A and $*$ is the Hodge star operator. For any A we also have the *Bianchi identity* $d_A F_A = 0$. In local coordinates we have:

$$F_{\mu\nu} = \partial_\mu A_\nu - \partial_\nu A_\mu + i[A_\mu, A_\nu], \text{ and}$$
$$L = Tr(F_{\mu\nu} F^{\mu\nu}).$$

Then the Yang-Mills equations become

$$\partial^\mu F_{\mu\nu} + i[A^\mu, F_{\mu\nu}] = 0.$$

year   Unit of time, $a = 31\ 556\ 952$ s. The year is not commensurable with the day and not

Z

Zeeman effect   The splitting or shift of spectral lines due to the presence of an external magnetic field.

zigzag phenomenon   Successive directions are orthogonal, causing Cauchy's steepest ascent to converge slowly. The successive orthogonality comes from optimizing the (scalar) step size with (exact) line search, but the issue runs deeper. In general, the zigzag phenomenon causes small steps around a ridge.

Zimm-Rouse model   A stochastic mathematical model for the dynamics of an idealized polymer chain. The linear chain is modeled as a collection of N beads connected by $N - 1$ springs. A system of stochastic differential equations for the N particles defines a multivariate Gaussian process. The equations can be solved using normal mode method. This model is a generalization of the *Ornstein-Uhlenbeck process*.

Zorn's lemma   If S is a partially ordered set such that each totally ordered subset has an upper bound in S, then S has a maximal element. Partially ordered means that for some pairs $x, y \in S$ there is an ordering relation $x \leq y$; totally ordered means for each pair $x, y \in S$ one either has $x \leq y$ or $y \leq x$. The Zorn lemma is equivalent to the axiom of choice.